CONTESTED CATCH
LOBSTER, LOCALISM, AND CANADA'S ATLANTIC COAST, 1870–1970

Suzanne Morton

Contested Catch: Lobster, Localism, and Canada's Atlantic Coast, 1870–1970 explores the complex development of the lobster fishery in Canada, focusing on the interplay between the ecology of lobsters, local fishing practices, evolving technologies, changing markets, and the role of the state. Drawing on nearly thirty different archives, this book spans the century from the expansion of the transnational commercial lobster industry and the introduction of government regulations around 1870 to the establishment of the first restricted-access ocean fishery on Canada's East Coast in the late 1960s.

Suzanne Morton argues that lobster regulation was always about more than just protecting the lobster population – it reflected deeper local, political, and economic forces at play. She examines how the Canadian state, keeping its enforcement budget to a minimum and wanting to avoid electoral reprisals, interacted with local communities, businesses, and political groups to regulate fisheries. While the government and other officials implemented formal regulations and turned to commissions and government science, local fishers operated with informal systems based on tradition, economic interests, and sometimes coercion.

Following the Second World War, with economists playing a prominent role, policy shifted from managing lobster as a sustainable resource to increasing the standard of living for fishermen and their families through rationalized efficiency, ultimately limiting access to who could fish. Even then, the fishery was shaped by both formal government efforts and local, social dynamics.

SUZANNE MORTON taught Canadian history at McGill University between 1992 and 2025, specializing in gender, the state, and Atlantic Canada.

Contested Catch

Lobster, Localism, and Canada's Atlantic Coast, 1870–1970

SUZANNE MORTON

UNIVERSITY OF TORONTO PRESS
Toronto Buffalo London

ISBN 978-1-4875-7177-1 (cloth)   ISBN 978-1-4875-7243-3 (EPUB)
ISBN 978-1-4875-7182-5 (paper)   ISBN 978-1-4875-7223-5 (PDF)

Library and Archives Canada Cataloguing in Publication

Title: Contested catch : lobster, localism, and Canada's Atlantic Coast,
    1870–1970 / Suzanne Morton.
Names: Morton, Suzanne, 1961- author
Description: Includes bibliographical references and index.
Identifiers: Canadiana (print) 20250276135 | Canadiana (ebook) 2025027616X |
    ISBN 9781487571771 (cloth) | ISBN 9781487571825 (paper) |
    ISBN 9781487572235 (PDF) | ISBN 9781487572433 (EPUB)
Subjects: LCSH: Lobster fisheries—Government policy—Atlantic Provinces—
    History—19th century. | LCSH: Lobster fisheries—Government policy—
    Atlantic Provinces—History—20th century.
Classification: LCC SH380.2.C3 M67 2025 | DDC 639/.5409715—dc23

Cover design: Will Brown
Cover image: iStock.com/THEPALMER

We wish to acknowledge the land on which the University of Toronto Press
operates. This land is the traditional territory of the Wendat, the Anishnaabeg,
the Haudenosaunee, the Métis, and the Mississaugas of the Credit First Nation.

University of Toronto Press acknowledges the financial support of the
Government of Canada, the Canada Council for the Arts, and the Ontario Arts
Council, an agency of the Government of Ontario, for its publishing activities.

Canada Council   Conseil des Arts
for the Arts   du Canada

ONTARIO ARTS COUNCIL
CONSEIL DES ARTS DE L'ONTARIO
an Ontario government agency
un organisme du gouvernement de l'Ontario

Funded by the   Financé par le
Government   gouvernement
of Canada   du Canada

Canada

MIX
Paper | Supporting
responsible forestry
FSC   FSC® C103567

Contents

Figures

Tables

Acknowledgments

Book projects are solitary enterprises only possible with the support, labour, and generosity of others. A standard research grant from the Social Sciences and Humanities Research Council of Canada financed archival research and allowed me to hire student research assistants. I would like to thank David Bent, Kaitlin Findlay, Colin Grittner, Adam Kuplowsky, Laurin Liu, Liam Mather, Carolynn McNally, Amanda Ricci, Sara Spike, and Haley Welch. These researchers all brought their individual talents and experience to their work. It was extraordinary, for example, to have Hansard notes interpreted by a former Member of Parliament.

I am also very grateful to the professional and volunteer staff of local museums and archives. This project depended on them a great deal. Linda Glass and the volunteers of the North Shore Archives (Tatamagouche, NS) and Barry MacKenzie of the Antigonish Heritage Museum were especially helpful. If you ever get the opportunity to visit the Argyle Township Court House and Archives in Tusket, NS, do it. It is a model of what local communities can accomplish. Other local archives I wish to acknowledge are the Cape Sable Historical Society (Barrington, NS), Centre d'études acadiennes Anselme-Chiasson (Moncton, NB), Fisheries Museum of the Atlantic (Lunenburg, NS), Maine Historical Society (Portland, ME), Maine Maritime Museum (Bath, ME), Massachusetts Historical Society (Boston, MA), Musée des Acadiens des Pubnicos et Centre de recherche (Centre de recherche Père Clarence-J. d'Entremont (West Pubnico, NS), Northumberland Fisheries Museum (Pictou, NS), Shelburne County Archives and Genealogical Society (now Barrington, NS), Thomas Raddall Research Centre and Queens County Museum (Liverpool, NS), Wallace and Area Museum (Wallace, NS), and Yarmouth County Museum and Archives (Yarmouth, NS). The Maritime History Archives (Memorial University of Newfoundland, St.

John's, NFL), McGill University Archives (Montreal, QC), and University Archives and Special Collections (University of Prince Edward Island, Charlottetown, PEI) also held useful fonds.

This project was based heavily on Canadian government archives such as Library Archives Canada (Ottawa), Nova Scotia Archives (Halifax, NS), Provincial Archives of New Brunswick (Fredericton, NB), Public Archives and Records of Prince Edward Island (Charlottetown, PEI), and The Rooms Archives (Newfoundland and Labrador, St. John's, NFL). In the United States I visited the Maine State Archives (Augusta, ME), National Archives at College Park, Maryland, and National Archives at Boston (Waltham, MA).

Generous colleagues, friends, and neighbours shared their knowledge of the lobster industry and kept their eyes open for relevant material. I am grateful to the late Boyde Beck, Walter Bethell, Jill Campbell-Miller, Lynn Farrell, Gerry Hallowell, Ed MacDonald, the late Mike Muise, Philip Slayton, Abel Stevens, Heather and John Stevens, Jamie Stevens, and Calvin Trillin.

At McGill I had the benefit of over thirty happy years with supportive and engaged colleagues in the Department of History and Classical Studies and undergraduate and graduate students who were almost always a real pleasure to teach. This was a remarkable privilege, and I am pleased to acknowledge here how much my colleagues and students gave me. This is also an opportunity to mention in appreciation someone who was both – Jarrett Rudy. I will always miss him. In terms of institutional support, I would also like to draw attention to the patient and overworked staff at McGill's IT support services and Interlibrary Loan who saved me more times than I can recount.

The manuscript benefited from the feedback of generous audiences at scholarly conferences. Most importantly, Tina Loo and David Meren – two very smart and very busy people — read initial drafts and the helped with focusing its themes. Advanced drafts were improved immeasurably by the intelligence and knowledge of Don Nerbas, Ed MacDonald, John Reid, and Peter Twohig. Dave Black reminded me that "localism and particularism" were the central themes of Alan Wilson's Atlantic Canadian history course at Trent in the early 1980s. While I offer credit to them for their insights, they may not agree with my interpretations, and I remain solely responsible for errors. I also benefited from the artistic talents of Marieke deRoos and Julius Reque and the cartography skills of Geoff Wallace and am grateful for a research grant from the McGill Institute for the Study of Canada that covered these costs.

Two of the most important imaginary readers I had in my head throughout this process did not live to see this project's completion. Bill

Parenteau was an exemplary human and scholar. Atlantic Canadian history is much poorer for his loss. He was always so supportive of this project, especially when I was not. I was also inspired by Mike Muise, lobster fisherman extraordinaire and admired his immense knowledge of the fishery, its politics, the ocean, and weather. He would not have agreed with everything I wrote, but we both would have enjoyed the discussion; I miss them both.

Other good friends also did not see this book to the finish line, and I would like to mention them, as lobster played a fundamental role in our friendships. Sarah Bradshaw encouraged me as a MA student the first time I had the nerve to put a live lobster in a pot of boiling water. Later, Rachel Summers, who loved to eat lobsters but whose Buddhist beliefs meant she did not kill them, willingly stood by the stove in my kitchen. Their premature deaths are an ongoing loss to me, Sarah's family, and their many, many dear friends. There will always be absent places around my dining room table.

It is fitting to finish acknowledgments with gratitude for the hospitality and generosity of friends on both this project and especially while caring for my late beloved parents. In particular, I would like to thank Lesley Barnes, Sandra den Otter and Simon Moore, Catherine Desbarats, Nancy Forestell and Peter McInnis, Elizabeth Kirkland, Catherine LeGrand, Brian Lewis, Tina Loo, Laura Madokoro, David Meren, Laila Parsons and Rob Wisnovsky, Daniel Woolf, and Julie Gordon-Woolf. The village of Port Medway stimulated my initial interest in the topic, sustained it, and continues to teach me that the fishery was, and remains, complex.

Suzanne Morton
Port Medway, NS/L'ketuk, Mi'kma'ki

Abbreviations

CMHA	Canadian Museum of History Archives
DFO	Department of Fisheries and Oceans
DM	Deputy Minister
FCS	Fish Canners Section
FO	Fisheries & Oceans
FPS	Fisheries Protective Service
FUNS	Fishermen's Union of Nova Scotia
FSC	Food, social, and ceremonial fishery
LAC	Library Archives Canada
M&F	Marine & Fisheries
MLA	Member of Legislative Assembly
MP	Member of Parliament
NFM	Northumberland Fisheries Museum
NSA	Nova Scotia Archives
PANL	Provincial Archives of Newfoundland and Labrador
PARO	Prince Edward Island Public Archives and Record Office
SFXUA	St. Francis Xavier University Archives

Introduction: Why Lobsters?

The year 2023 marked the 150th anniversary of regulations governing one of Canada's oldest, continuously state-managed fisheries. It was not surprising that this anniversary passed uncelebrated, as regulation of this fishery has been contentious from the beginning. When commercially fished, lobsters are always a contested catch. Lobster suggests a kaleidoscope of clashing images – luxury eating, traditional fishing villages, dangerous work, and violent conflicts on both land and water. It is a fishery associated with myths and stereotypes but rarely understood as an industrial product, one made possible through mid-nineteenth-century innovations in preservation and transportation. Before industrialization, lobster was not an important commodity, but from the moment it had market value, it was entwined with the emerging Canadian state, politics, and the destruction of extractive capitalism.

Lobsters are and were widely distributed throughout the Canadian Atlantic coast. The importance of lobster is literally written on the landscape. The Canadian Geographic Names Database lists fifty-two places or features in New Brunswick, Newfoundland and Labrador, Nova Scotia, Prince Edward Island, and Quebec with lobster or homard in the name.[1] Today it is unquestionably the most important Canadian fishery – but its East Coast significance was apparent by the late nineteenth century. By the 1890s, lobster's commercial value was comparable, or in a couple of years slightly exceeded, the regional groundfish catch (a category that includes cod, haddock, pollock, and flounder). What was called the "lobster industry" employed many more fishermen on the water and a large number of men, women, and children in the processing sector. In a monumental study that shaped federal fishery policy in the postwar era, the author noted that based on pre-1939 numbers "eastern Canada as a whole was the land of the lobster, not the cod."[2]

The dispersed habitat of lobsters and the decentralized nature of processing meant that fishing and packing were important almost everywhere along the coast from the North Shore of Quebec to the border with Maine on the Bay of Fundy. Until the late 1930s, lobster was predominantly sold as a preserved canned product. Over time, there was a general shift from a canned to live market or frozen product, but the specific timing of the change varied greatly according to place. Geographic proximity and transportation access to the large urban American coastal markets meant that the southern Atlantic Coast and the Bay of Fundy developed a live market trade in the late nineteenth century. Distance and lack of infrastructure kept most of the important Gulf of St. Lawrence fishery shut out of this more lucrative market until the inter-war period.

Before the late 1960s, lobster was a poor people's fishery, a term that highlighted the relatively little capital necessary for individuals to pursue it. In the Gulf of St. Lawrence region until the Second World War, it was common for individual lobster fishermen to be tied to a specific canner through permanent debt obligations or be employed to fish for a processor in exchange for some combination of wage and company gear. Elsewhere, independent fishermen owned their own boats, fished their own bait, and made their own traps so capital outlay, at least in the nineteenth century, was largely restricted to rope and twine. In the twentieth century, capital expenditure increased with the advent of motorboats, mechanical haulers, and their use of fossil fuels primarily in the form of gasoline, but it was not until the 1970s with the advent of larger fibreglass boats, electronic equipment, synthetic ropes, and most importantly the increasing cost of a restricted access licence, that the fishery lost its status as the employment of last resort.

Unlike farmers and industrial workers, fishermen, with important exceptions, are virtually invisible in accounts of political history on the East Coast outside of Newfoundland.[3] In 1913, for example, there were 25,000 fishermen associated with the lobster fishery; roughly the same as the number of coal miners in the region at that time. It is striking how much more we know about miners.[4] Lobstermen were geographically spread out, rarely organized at anything beyond the community level, pursued the fishery on a part-time basis, and tended to be deeply entwined in regional partisan politics. Both the Liberal and Conservative parties, especially before the First World War, understood lobster fishermen's importance as voters and actively courted their electoral support. Canners too were also tied to both traditional parties, and government regulations and policies always reflected and attempted to balance the various interest groups who held sway.

We see glimpses of the nascent Canadian state in the early regulation of the lobster fishery, which was one of the first points of contact many East Coast fishermen had with the new country. Marine and fisheries regulations, along with tariffs, the post office, and prohibitions on smuggling were one of the relatively few ways that coastal communities encountered the federal government on a daily basis. Until after the Second World War the image presented of the state through fishery regulations was that it was weak and inconsistent. It lacked the capacity, legitimacy, and the authority to implement its regulations in the Maritimes and the Gulf of St. Lawrence. Conventions on the water and at the wharf were in fact a mixture of formal and informal practices. While the authority of the state could be dismissed, there were examples where local communities invoked the state's power to further their own agenda. The ongoing centrality of natural resources in the Maritimes meant, and to this day means, that the environment is always political. Conservation and economic development are posed as conflicting objectives, and it is striking to see nineteenth-century predecessors struggling with environmental sustainability. In the realm of resource conservation, saltwater fisheries are especially interesting as a federal jurisdiction, laying bare the obstacles of governance from a distance and the often competing local, provincial, and federal interests. Natural resources outside jurisdictions classified as territories fell under provincial control, and the exceptional federal claims to saltwater fisheries foreshadowed later conflicts over offshore gas and oil rights.[5]

Lobster regulations, in turn, help shaped views of the state in the region, and it is not difficult to imagine agreement with anti-Confederation politician Joseph Howe's 1867 prediction that Confederation would force the region to "submit to the dictation of those who live above the tide, and who will know little of and care less for our interests and our experience."[6] Atlantic Canadians today might characterize this as being "out of touch." Of course, Indigenous peoples encountered a very different state that was willing to use both direct and indirect coercion to achieve its aims. There were no immediate electoral political consequences when dealing with Indigenous communities.[7]

As a new commercial fishery that was born in the industrial age, the lobster fishery did not have a set of established pre-capitalist traditions to draw upon and so emergent community practices developed in tandem and were negotiated with the Dominion of Canada and new transnational corporate networks. There were no specific historical customs to turn to for understandings of individual and community rights or sustainability, and so emerging market forces, state regulations, and conventions borrowed from other species shaped fishing practices and attitudes towards conservation.[8] Different conceptions of the commons were held by individuals, local communities, businesses and governments, each of whom regarded themselves as resource managers.

Underlying all discussion and concern regarding lobster conservation were complex and sometimes contradictory notions about lobstermen's behaviour and motivations. Almost everyone believed that intensive overfishing posed real dangers to the lobster stock, but there was no agreement about who was responsible and how best to stop it. At times in the nineteenth century, packers, crudely characterized as greedy Americans, were held accountable. In 1887 PEI government Fishing Inspector J. Hunter-Duvar cast the problem as insatiable canners and urged state action so that "more restraint ... be put on the mercantile adventure."[9]

More often than not, however, the blame was placed squarely on the impoverished fishermen. Long before Garrett Hardin published his influential and controversial 1968 article "The Tragedy of the Commons," bureaucrats, politicians, and community-based experts argued that long-term stock sustainability was sacrificed for immediate profits by shortsighted and self-interested fishermen.[10] These land-based actors imagined the fishery as a completely open resource commons where the absence of exclusive rights of access and ownership meant constant competitive exploitation of the resource. In this framing, fishermen had no long-term interest in protecting the stock. They thought only of immediate gains and needs and were not restricted by informal regulations. In an 1896 letter to a Member of Parliament concerning lobster regulations, the top federal fisheries bureaucrat wrote of "the general indifference prevalent respecting the future welfare of the fisheries" among fishermen. "So long as any fish can be caught, alarm as in the future appears rarely to arise."[11] In the same vein cooperative organizer J.T. Croteau, who held the Carnegie Foundation chair in economics and sociology at St. Dunstan's College in Charlottetown from 1933 to 1946, concluded that

the whole idea of saving and foresight is foreign to fishermen as a group. There are exceptions; but on the whole, fishermen live in the present. They concentrate on catching what they can, whenever, they can. Conservation measures in the lobster industry, the Island's richest fishery, can be enforced only by the most extensive government patrol. If left to themselves, the fisherman would fish out the lobster in five years.[12]

The description of PEI lobster fishermen as "irresponsible and improvident" echoed politician J.J. Tucker's 1903 condemnation of Magdalen

Islands lobstermen who "with the usual want of forethought, which is a distinguishing characteristic of most fishermen ... [they] look only to the momentary gain to be derived from immediate large catches, and consequently look with little favour upon the efforts of the Government to perpetuate the fishery on their behalf."[13] This widespread perspective adopted as "common sense" completely overlooked that the marine ocean commons was never completely open. As this book will demonstrate, informal property rights were also always a factor and, in practice, fishing grounds combined informal individual and community customs with formal regulations.[14] Over thirty years ago British historian E.P. Thompson reminded us that "custom is local," and we see in the pages that follow that communities often attempted to regulate access to the common resource of lobsters themselves.[15]

Before the late 1960s, lobster did not fit neatly either as a state-managed commodity nor as a communally managed natural resource. It was both, with the added complexity of certain access rights strongly associated with individuals or families. These incongruent strategies were held concurrently, and this resulted in conflict. For nearly the first hundred years of its existence, the commercial lobster fishery was a modified open commons. Anyone could fish, but fishermen were subject to community mores and an ever-changing series of state regulations, both imposed and requested, whose objectives were a mixture of conservation, price stabilization through supply-side management, and to aid the enforcement of the regulations themselves. Fishermen, according to their involvement as independent commodity producers or as employees, held a range of views. The independent commodity producers were, as historian Brian Payne points out, both "stewards and capitalists," while at the same time those employed or indebted to canners were in situations more akin to feudal servants to packing masters.[16] Payne's description of the independent commodity producers closely mirrors the non-state sustainability tradition usually associated with economist Elinor Ostrom who argued that common-pool resources were best safeguarded through community-based customs and self-regulation. She held that more than distant bureaucrats, local community members held relevant local environmental knowledge. Moreover, by excluding outside competitors, they could hold local people accountable by punishing those who transgressed accepted practices.[17] So, from the beginning of the industry, born out of capitalism, there were fishermen and packers who rejected the notion of the commons and held that the emerging market needed to be facilitated, managed, nurtured, and regulated by either capital or the state. There were also local fishermen, who transferred their world view of the communally managed commons and rejected this privatization or the distant state management of the sector by those who were not connected to local practices or knowledge. The understanding of the commons was not commonly held.

Contested Catch: Lobster, Localism, and Canada's Atlantic Coast, 1870–1970 argues that the development and regulation of the lobster fishery were shaped by the habits of the lobster itself, changing technologies, markets, and ideas about the state. Most importantly, the fishery's development and regulation were fundamentally formed by local and social relations. Local contexts provided meaning, complexity, mutability, and limits to state power and in turn tell us not only about the lobster fishery but also a great deal about the state at both the local and federal level. The book covers the period from the dramatic expansion of the commercial industry around 1870 until the late 1960s when the lobster fishery became the first significant restricted access ocean fishery on Canada's East Coast.

As a fishery that had a wide geographic range, the concentration is primarily on the Maritimes and Quebec and to a lesser extent the more limited lobster fishery of Newfoundland. Before 1949 Newfoundland was an independent colony and outside the jurisdiction of the Canadian federal state. Its lobster fishery nonetheless shared much in common with Canada's Atlantic coast through markets, technology, and science. Maritime merchants and American companies were behind the canning boom that peaked there in 1889. Although first international diplomacy and subsequently a greatly depleted stock created a different canning structure than the western side of the Gulf of St. Lawrence, its late integration into the American live market coincided with that of its geographic neighbours.[18] With the focus on the local, this study offers less attention to the consumption side of the industry, which was primarily international and far away from where lobsters were landed. Brian Payne's recent book *Eating the Ocean: Seafood and Consumer Culture in Canada* opens a new avenue in the Canadian literature as it frames fish as a food rather than primarily a natural resource.[19]

The lobster industry itself was complex. It was decentralized and required very little capital to enter and yet from the very moment there was a market for canned lobster, there was also the keen awareness that the resource could be depleted and disappear. Its origins not only coincided with the emergence of the Canadian state around 1870 but also in the context where overfishing had already reduced the forerunning New England lobster fishery.[20] As American canners moved north, conservation concerns existed at the very moment the commercial fishery was born. Debates about state regulation became more vigorous in the 1880s when it appeared as if the Canadian fishery would follow the United States' example and weaken the stock. The regulatory aspirations of the various federal departments responsible for fisheries greatly exceeded their enforcement capacity until the 1950s.

From the 1870s, the new, weak, parsimonious, and distant Canadian state had limited capacity to enforce its regulations and the issue of rights and regulations – both formal and informal – were always in dialogue. The Canadian state was also in constant negotiation with businesses about who could regulate best the lobster industry and how local political parties could accrue immediate benefits. Though the commodity and business were transnational, local conditions and local market forces shaped understandings about the right of access to the fishery and what regulations were appropriate. While politicians, bureaucrats, scientists, and packers issued formal prohibitions and regulations, gathered information, and launched educational campaigns, those who fished had parallel informal systems of regulating the fishery based on traditional practices, economic power, and at times physical coercion. After the Second World War, the focus on conservation, or rather the management of the lobster fishery, was supplemented by a concern about conserving coastal communities as economists; first Keynesian and later neoliberals took the leading role in fishery policy formation. The fishery that developed reflected both these formal and informal forces and the divergent agendas of those connected to the Atlantic coast's most important fishery. The regulation of lobster was always about more than protecting crustaceans.

Since there was no significant commercial market for lobsters before 1867, unlike salmon for example, Nova Scotia and New Brunswick had no pre-existing regulations. The entry of Prince Edward Island into Confederation in 1873 coincided with the first federal lobster fishing statutes. In this way, the new federal government simultaneously established regulation over a fishery and oversaw the invention of "Canadian" interests as counterposed to "American" companies. By the 1880s, the phrase "Canadian lobster" had meaning as the new nation established dominion over its natural resources.[21] In a context of residual anti-Confederation sentiment and repeal, an emergent and fragile nationalism was fanned as American business interests were portrayed as exploiting "Canadian" resources with fears that abundant local stocks would be depleted by outsiders. From the late 1880s onward, after the completion of the transcontinental railway, lobsters became part of a nation-building activity as ecological nationalism with at least six (unsuccessful) attempts to introduce Canadian Atlantic lobsters to the Pacific waters off British Columbia. This was building Canada sea

to sea in the most literal sense. The regulation of the lobster industry
was also a means to dispense the dividends of the state through partisan patronage in the ensuing jobs and stipends of overseers, wardens,
and guardians. Many politicians themselves were hardly disinterested.
There was a remarkable overlap between regional politicians and lobster canners. Packing was something that many Members of Parliament
and senators had their hands in.[22]

In this book, I have chosen to use the traditional nineteenth- and twen-
tieth-century term "fisherman" rather than adopt the gender-neutral
word "fisher." There is an ongoing debate about appropriate terminology, and my choice may have been different if this work focused on the
present day when some women fish lobsters. Fisherman is useful when
exploring a specific historical context that almost exclusively excluded
women, as it reinforces the importance of masculinity to the lobster
fishery and accompanying ideas about providing for one's family, independence, and a patriarchal order.[23] Unlike the inshore fin-fish fishery,
lobster was less likely to draw upon household production, although
different members of the household might be involved in industrial
processing.

The sources behind this book are overwhelmingly based on government material and the publications of the fishery industry. Within these
voluminous papers (especially for the pre-First World War period), it is
rare to see the candid opinions of scientists and bureaucrats, although
politicians remained remarkably forthright. The voices of fishermen
appear in their letters to Ottawa, and they are recorded as witnesses
before government commissions and inquiries, but those who worked
on the water or in the factories are often invisible and anonymous. I
located one remarkable diary by a Shelburne County lobster fisherman
that was taken over by his wife in April 1898 after he drowned while
tending his traps.[24] A handful of surviving canners' account books give
us some view of payrolls.[25] Packers and the commercial fishing sector
are much more visible in their active lobbying of Ottawa through correspondence with politicians and their organized petition campaigns.
From 1914 the commercial fisheries industry sponsored a monthly journal, *Canadian Fisherman*, and the lobster packing industry was highly
visible in its pages through its first thirty years. Much more of this
perspective was provided by some of the archived records and scrap-
books of Richard H. Williams, who served as the Canadian manager of
Roberts, Simpson and Company of Liverpool, England, and Halifax,
Nova Scotia. In the early 1900s this company was the largest exporter
of canned lobster in the world, and Williams considered himself an
authority on both the lobster industry and its history.[26]

Archival research in Ottawa, Washington, Boston, and throughout Maine and Atlantic Canada shaped and reshaped the direction of this project. It convinced me that it was impossible to understand the history of the politics around regulation without considering the economic structure of the industry, its transnational nature, and its technology. While the basic gear remained frozen by regulations made at the turn of the twentieth century, fishing practices were intensified by the motorboat, mechanical hauler, capital intensive canning technologies and freezing, and the industrial consolidation made possible through paving rural roads and trucking.

This book is located at the intersection of several historical themes. Although the focus is on the social politics of regulation, it also serves as the first general history of the "Canadian" lobster fishery and contributes to ongoing historical conversations on natural resources and political economy, building on previous studies of formal and informal practices and strategies.[27] There is a rich scholarship on the Maine lobster industry, and although there are important similarities, the history of regulation is remarkably different as a result of local jurisdictional responsibility and the relative absence of packers.[28] The larger history of the Canadian lobster fishery is complex, but it has been explored in wonderful local histories where the particularities of location and environment offer coherence.[29] The lobster fishery has also appeared as an important aspect of Acadian community studies.[30] While there is a strong emphasis on the local, this book does not provide community social histories of the permanent and transient seasonal labour force of men, women, and children who processed lobster on land. General histories of the Atlantic fishery have disproportionately focused, mirroring government policy, on the offshore and fin-fish fisheries. This literature has emphasized industrialization and centralization and has overlooked an important fishery that did not share these traits.[31]

There is a substantial history of fisheries management and government-sponsored fisheries science, but this book bridges centralized institutional work with what was happening on the water, wharves, and in local communities. It tries to link the Department of Fisheries, in its various incarnations, with politicians, industry, and fishermen.[32] Much of the related recent scholarship can be linked to either the collapse of the cod fishery and the 1992 moratorium or ongoing issues of Indigenous people's access to the fishery and treaty obligations. This book's conclusion, which focuses on the post-1970 period, contributes to understanding both important contemporary issues.

There is an incomplete but strong literature on Canadian state formation and resource politics often centring utilitarian conservation. An older body of work on natural resources, such as H.V. Nelles' important 1974 *The Politics of Development: Forests, Mines and Hydro-Electric Power in Ontario, 1849–1941*, drew on business, economic, and political history. More recent scholarship has used environmental history, politics, and economic forces to highlight specific dynamics such as the competing interests of power and Pacific salmon or the challenges of draining prairie wetlands.[33] The theme of state formation and resource politics is the focus of Stéphane Castonguay's valuable study of how through science, natural resources became a tool for modernizing state power in Quebec.[34] The limits of these same powers is explored by Tina Loo, who through the examination of wildlife conservation demonstrated (like this book) that rural resource users, community networks, and state authority coexisted and it was not a simple trajectory moving from local folk custom to state-directed scientific management.[35]

Like resource regulation, much contemporary scholarship on the state and natural resources focuses on the complex and vital issue of natural resources, Indigenous peoples, and the related politics of justice or erasure.[36] While the scholarly literature related to this book's concerns have broadly moved from political economy to colonial extraction, this book attempts to keep its feet in both traditions, highlighting the formal and informal local political responses. National narratives are combined with conflicts that occurred in particular places, a theme often not emphasized in non-Indigenous scholarship.[37]

Linked to resource extraction, *Contested Catch* also contributes to scholarship on how one commodity played a role in the transition from a rural pre-industrial economy to a rural capitalist and industrialized one. This process is associated with labour mobility (but not always rural depopulation), occupational pluralism, commodification, and resource extraction. Following R.W. Sandwell in her *Canada's Rural Majority*, this book focuses on rural people who supported themselves participating in all kinds of different economic opportunities according to the vagaries of commodity booms and busts, season, age, gender, and location. Those involved in the lobster industry were independent commodity producers, straight wage earners, and employer-bound piece workers, or perhaps some combination of the above. To paraphrase Sandwell, they were not fully capitalist, nor proletarian, nor peasants.[38] Similarly, historian James Murton, while discussing the commodification of nineteenth-century natural resources in general, might have been addressing lobster when he observed that "as market-based systems of production and work arose, environmental problems arose to bedevil policymakers and local users of nature."[39]

The book begins with a chapter looking at lobsters themselves and the human predators who tried to catch them. Lobsters and the related fishing technologies spawned the subsequent formal and informal regulations. The second chapter "Lobster Becomes a Commodity," explores the emergence of lobster as a new product and the parallel development of a lobster industry. The economic geography of the industry combined with the ecology of lobster to create a labour-intensive, decentralized fishery. Essential to understanding the development of the live and canned sectors is the state regulation discussed in chapter 3, "Regulating the Lobster Fishery." Regulation was the way many people in Atlantic Canada encountered the new Canadian state, but the state they encountered was weak – so inchoate that informal parallel community practices evolved beside formal regulations.

The incipient and usually sparse state looked for non-coercive solutions (hatcheries) and hoped knowledge would foster good policy. Chapter 4, "Politics of Knowledge: The Lobster Question," investigates the "lobster question" as it emerged as a political issue. The focus is on the creation of political knowledge both through commissions and inquiries and through government-sponsored science that before the Second World War tended to be practical and empirical and therefore offered space for local or situated knowledge. Although most commissions and inquiries before 1930 travelled to local communities, knowledge was also disseminated from Ottawa and shaped by its political agenda. The location of hatcheries was all about extracting political gain and votes, and the Pacific lobster experiments were moulded by the politics of region within the federal government.

Both chapters 5 and 6 look at politics. Chapter 5, "Rejecting Regulation: Direct Action, Force, and New Strategies," focuses on informal politics and direct action and argues that notwithstanding the space for local knowledge, the state's attempt to regulate the fishery was neither accepted nor effective. It looks at how and why enforcement was limited and by what means locals and packers could use the state to do their bidding, even as they evaded fisheries regulations. The sixth chapter, "The Politics of Lobster and Catching Votes," turns to formal partisan politics and patronage. It argues that especially before the First World War the lobster fishery was at the centre of local traditional electoral politics. Packers and fishermen organized formally to put their specific interests forward. Politicians were very aware that lobstermen's votes mattered. These first six chapters show how economic, political, geographic, and cultural forces within the region combined with the particular nature of the lobster fishery (especially its low capital requirements) and led to the fishery being governed largely by an informal social mandate. While this mandate is mediated by the state, it is not fully controlled by it.

The final chapter, "Lobster Regulations as Social Policy," looks at the way postwar federal lobster policy was less about managing an economic resource and more about regulating fishermen and their families. Lobster fishery regulations had been used as a pre-Keynesian form of state welfare to provide relief in times of distress. After the Second World War, while the federal government focused on a modern industrial offshore fishery, the inshore fishery anchored around lobster was transformed by Keynesian economists who sought to improve fishermen's standard of living through efficiency. The emphasis on economic governance with the parallel consolidation of the welfare state had the unintentional consequence of serving a local social mandate that had been delivered in the past by the state's tolerance of a type of "moral economy." Another unintended consequence was that idealist bureaucrats and academic experts opened the door for neoliberal economists, the end of the commons, and the privatization of a natural resource.

Although *Contested Catch's* archival research ends with formal closure of the lobster fishery commons for most of the region by 1970, the conclusion makes a narrative connection between the new limited entry fishery and the present, with particular emphasis after the 1990s on Indigenous issues. I hope it becomes clear as to why the book's pre-1970 findings remain important today. The decision to title the book *Contested Catch* was intended to strike a resonant chord with the present-day fishery and provide an historic context that emphasizes that formal and informal regulations around lobster have always been contentious.

I also hope that the reader might come away from this book also thinking a bit differently about how most history has been written regarding industrial capitalism and the Atlantic region. What becomes Atlantic Canada is not exclusively the passive result of external forces, such as metropolitanism, dependency, and underdevelopment. Fred Burrill has already noted that histories of rural capitalism and the fisheries "call the underdevelopment approach into question on a number of levels."[40] Historians' past preoccupation with heavy industry and manufacturing in the post-1870 period have often ignored natural resources, except for the capital and labour-intensive activities around coal and wood pulp, and lead us to underplay how local factors such as occupational pluralism, economic decline, proximity to outside markets, local conceptions of place, the operation of party politics, and deep-rooted traditions of direct action have also created a distinct present.[41] These local conditions combined with what Burrill reminds us are "the central insights of underdevelopment theory that the fate of the Maritimes has always been tied up in a world-scale struggle over resources and that this struggle has manifested itself in the political structures governing the region." The industrialized lobster fishery provides a concrete example of this.[42]

Exploring the lobster fishery offers insights into the difficulty that Ottawa had in establishing distant control over regional resources as federal law conflicted with concurrently emerging local customs and practices. It was one of the many places where the state and its citizens negotiated their relationship. Federal regulation could represent the state as a source of charity and patronage, or as a distant authority asserting its coercive power – sometimes through force, other times through indifference. Though these dynamics remained contested, federal lobster regulations continue to shape local communities to this day.

Chapter One

Lobsters and Those Who Caught Them

Humans living along saltwater coasts have always caught and eaten lobsters when freshly available. Although lobsters are easily caught when they move close to shore seasonally, they do not preserve well using traditional methods such as smoking, drying, or salting. To minimize severe food poisoning, lobsters are usually cooked alive and consumed immediately. The culprit is a saltwater Vibrio bacterium (the freshwater version is *V. cholerae*, the source of cholera), which occurs naturally in the coastal waters of Canada and the United States, in fish, and in shellfish. Vibrio multiplies rapidly in the hours after a North American lobster or *Homarus americanus*'s death.[1] The biology and ecology of lobsters (and the presence of Vibrio bacteria) set the context for *Contested Catch*, as this chapter explores lobsters, and the strategies and technologies humans used to catch them.

Lobsters belong to the family of Arthropoda, which are among the oldest beings on the earth and compose approximately 80 per cent of all animal species today. Two important families of edible lobsters exist; the cold water Nephorpidae composed of the larger *Homarus americanus*, European *Homarus Gammarus*, and *Nephrops norvegicus* (Norwegian lobster, scampi, or langoustines) and the tropical or subtropical *Palinuridae*, or spiny or rock lobsters of the Caribbean, South Africa, Asia, and Australasia. North American lobsters rarely faced direct consumer competition from anything but canned spiny lobsters.[2]

Homarus americanus, the subject of this book, are relatively sedentary, long-lived, nocturnal, bottom-dwelling, crawling creatures with seasonal migration patterns connected to water temperature. Inshore (or coastal) lobsters generally remain quite local, not moving more than a few kilometres offshore in winter to avoid extreme cold. Mature offshore lobsters move longer distances towards the shore in the summer and averaged a range of thirty-two kilometres, with most having a small territory and an exceptional number moving much greater distances.[3] The concentration of lobsters vary year to year, the result of food supplies, water temperatures, and the presence of predators. Adult lobsters' greatest predators are humans, but at different stages – larvae, juvenile, and adult – they are vulnerable to cod, Atlantic striped bass, wolfish, cusk, longhorn sculpin, haddock, eel, and various kinds of hake. Jonah and rock crab are recognized predators of larvae, and the expanding number of green and Asian shore crabs is a present-day threat.[4] All crustaceans are vulnerable to climate change, with their ability to thrive increasingly threatened by ocean acidification, warming water, and changes in the availability of diverse food sources. Ocean temperatures influence growth and fertility, leading to significant differences between the size at maturity between those living in the more constant temperatures of the Bay of Fundy or the Atlantic and those in the more variable temperature extremes of the Gulf of St. Lawrence.[5]

The growth of a lobster is determined by the number of times it moults; with each shedding of the shell, its size increases. This can happen multiple times a year when it is young and less frequently with maturity at six to seven years. After a moult, usually when the water is the warmest, a lobster is said to be "soft shelled," which has an impact on the live fishery, as its shell is neither full of meat nor strong enough to survive shipping. The soft-shell period is especially important for females, as it is the only time mating can take place. With successful fertilization, eggs (8,000 eggs on a one-pound lobster, while a nine-pound lobster might carry 100,000 eggs) are carried externally on the female "berried" body. When the eggs reach maturity after nine to twelve months, they are released as larvae, floating on the ocean where it is thought that only one in a thousand survive to the juvenile stage and sinks to the bottom.[6]

Archaeological evidence demonstrates that before commodification, lobsters were eaten locally. Lobsters comprised a small but important part of Beothuk and Mi'kmaq seasonal diets and were and continue to be used by the Mi'kmaq as fertilizer, bait, totem, tobacco pouches, and pipes.[7] Though less extensively than other marine invertebrates, lobster tails were cured through drying and smoking.[8] As an abundant food source, not easily preserved, lobsters also comprised part of the seasonal diet of early European settlers who fished along the coast.[9]

In the mid-nineteenth century, lobsters could be found in the cold saltwater from North Carolina to Labrador – all along the Atlantic coast, although they were much more plentiful in some districts than others. There is no question about the abundance of lobsters before what fisheries officials referred to as "indiscriminate" fishing. Moses

H. Perley, in his 1849 *Reports on the Sea and River Fisheries of New Brunswick*, called attention to the "extraordinary numbers" present in the Baie des Chaleurs and Shippegan Peninsula where every potato hill was supplemented with two or three dead lobsters for compost. After the infamous August Gale of 1873, a Prince Edward Island North Shore canner reported that the lobsters driven ashore formed a row from 0.3 to 1.5 metres deep and that there was an average of a thousand of them to every 3 metres of shoreline.[10] This almost unimaginable abundance did not last. It is remarkable, therefore, that unlike other fish, from the moment of its emergence as a mass commodity in Canada, lobster was labelled as a fishery at risk. In 1887 a Canadian Lobster and Oyster Commission decried overfishing in Maine and Massachusetts and warned Canadians not to follow the American example, for if they did not take immediate action lobsters would "soon become extinct."[11]

As lobsters usually do not migrate very far and are relatively slow to reach sexual maturity, any intensive fishing almost immediately diminished local stocks. This local context is important and must always be kept in mind. In 1874 W.H. Venning, Fisheries Inspector for Nova Scotia and New Brunswick, noted that American lobster packers had "fished out" their own waters and moved to Canada after the small size of the remaining lobsters in Maine rendered "further operations unprofitable." He collected data demonstrating that the size of Canadian lobsters had dropped already from the three to five pounds recorded when an area was first fished to a pound and a half as the number of factories in an area increased.[12]

The collapse of the lobster stock was noted as early as 1871, when Venning had warned of "exhaustion from fishing" and the need to regulate "the rapidly increasing business."[13] Canners responded by packing more smaller-sized lobsters, a labour-intensive and therefore more expensive process. In 1874, Prince Edward Island canners used three and a half lobsters to fill a one pound can; by 1884, this took five lobsters, and by 1892, six or seven.[14] By 1909 W.F. Tidmarsh of the Portland Packing Company wrote to the federal Marine and Fisheries Committee, claiming that should a size limit of between six and eight and a half inches be introduced and actually enforced, every cannery in Prince Edward Island would close.[15] There were no more large lobsters available. The report of a Prince Edward Island cannery manager saying that lobsters "as large as a grasshopper" were being canned and a similar statement in the St. John's, Newfoundland, *Trade Review* that thirty or forty fish were needed to fill a one pound can seems unlikely in terms of the labour needed to extract the meat, but both claims were reported in a 1911 American study.[16]

Lobster had been fished in shallow waters in a variety of methods. "Indian-fashion" (acknowledging a pre-settler and ongoing Indigenous fishery) lobsters were caught at night by torchlight with wooden spears with two prongs on either side that kept the lobster alive. Settlers fished during the day, throwing oil on water to cast a shadow and luring lobsters out from rocks with cod heads used as bait and then speared them with gaffs.[17] Although closed pots or creels – recognizable precursors to the post-1890 lobster trap – were used in the English lobster fishery, they were seldom used in the Maritimes, as they were "considered an expensive appliance."[18] Nova Scotians generally pursued a hoop fishery (sometimes called curley fishing) in which bait was placed in the centre of a net stretched over a two-foot iron ring or hoop that would be pulled up when a lobster went for the bait. A single fisherman in a dory might tend twenty of these hoops at a time. These "hand pots" caught many more lobsters than spearing and kept the lobster alive, which was important for safe canning. Hoop fishing persisted on a large scale on Nova Scotia's Eastern Shore until at least 1894.[19]

In some places in the Gaspé Peninsula, long lines of cod heads were strung together, and when lobsters came to the surface to feed, they were taken with dip nets. This method, which did not need the physical strength required to haul water-soaked heavy wood traps from the bottom was, according to one fishery official, "followed mainly by old men and children" so "no serious harm" was caused by prohibiting it. It did not impact middle-aged men who were cast as important citizens and family breadwinners.[20] All fishing methods other than lobster traps became illegal in 1910, as they tended to catch very small lobsters. Cod head trawls remained in use in the Gaspésie and northern New Brunswick in 1912, even after these methods had been outlawed.[21] As the fishery expanded and the number of lobsters declined, hoop pots fell out of fashion as they were limited in the number that could be fished at one time. Hoop pots also undermined conservation efforts, as this technique necessitated the lobster being very close to shore, which usually coincided with the spawning and soft-shelled season of summer.[22]

Wooden traps or "cages" with laths were introduced to the Maritimes in the latter half of the 1860s by New England canners who extended their operations northward. Fishing intensified as more traps were used to catch fewer fish. In 1885 some fishery inspectors began reporting the number of traps in their districts, and this shortly became part of the official statistics collected by the Fisheries Branch. For example, in 1887 443,115 lobster traps were reported in use, and this number increased to 768,479 traps in 1892.[23] By the 1910–11 season this number had nearly doubled again, as there were a reported 1,504,872 traps in use, all to

Figure 1. Lobster Spearing by Torchlight in Canada. *The New Graphic*, 19 March 1881, 32.

maintain – roughly – the same annual pack. There were not only more traps, but the design of traps evolved so that they were more efficient and could catch more lobsters. Right through to the recent present, there was significant local variation in the preferred design used.

New models meant that it was more difficult for trapped lobsters to escape, and it was easier for individual fishermen to fish more traps in deeper water. Variations of the box traps, such as the Wheeler or "Patent," diamond, parlour, and jail traps could be tweaked with double, three, or four headers (entry funnels in mesh).[24] In the early 1890s, the two-sectioned trap was adopted by packers with the lobster first entering the baited "kitchen" through mesh doors, sideheads, or funnels, shaped so that it was easier to enter than exit. The lobster then crawled up the funnel and fell into a second compartment called the "bedroom" or "parlour" where it was trapped. The 1898 Royal Commission noted that the Wheeler or parlour trap was in use in "many localities" and promoted by canners. It was attributed to the invention of E.A. Wheeler of Botsford, New Brunswick (located between Port Elgin and Cape Tormentine), who appears to have been a young labourer working in a household that included fishermen.[25] The parlour trap with its two distinct rooms was so well known that it appeared in *Popular Mechanics* in 1906.[26] These wooden traps only lasted around three seasons and were

Figure 2. An example of hoop fishing. George Richardson Hind, "Lobster Fishing River John," 1876 from George Richardson Hind, *The Pictou Sketchbook*, 1876 watercolour and gouache on paper Collection of Art Windsor-Essex. Gift of Mr. and Mrs. Kenneth D. Heath, 1967 1967.010.031.A-B Original in colour.

CUTAWAY OF WOODEN LOBSTER TRAP

Bait Bag   Funnel

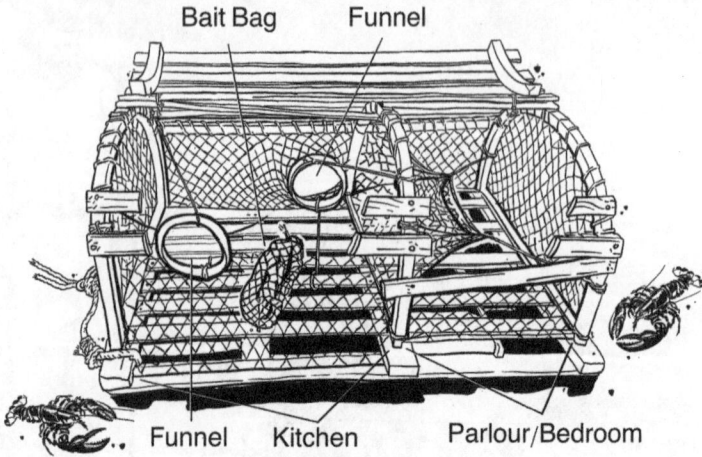

Funnel   Kitchen   Parlour/Bedroom

Figure 3. Since their introduction in the late nineteenth century, hooped wooden traps had many local variations in design and names for their different parts.

Illustration by Julius Reque

frequently destroyed by storms. At various times, different locales used them to fish differently either on single lines, a few traps on a trawl moved regularly, or the practice in the Northumberland Strait where a long trawl of a series of traps was stationary for the entire season.

Although the strategy for setting traps differed, the traps themselves were the same for most of the twentieth century. The parlour trap consisted of bent spruce or hemlock bows, laths, knitted twine mesh, weighted with flat stones, and baited with salted herring or cod heads.[27] Regulations froze the fishery in this basic technological form with the unintentional consequence that it was not subject to the same capitalization and industrialization as other sectors of the industry. After the Second World War, the Department of Fisheries promoted innovation with experiments with metal traps, space-saving collapsible domes, plastic, and folding traps but the lengthy close season of the Canadian lobster fishery meant that home-made wooden traps continued to be a rational and strategic choice for fishermen who often had more time than money.[28] Plastic coated wire traps were introduced around 1970

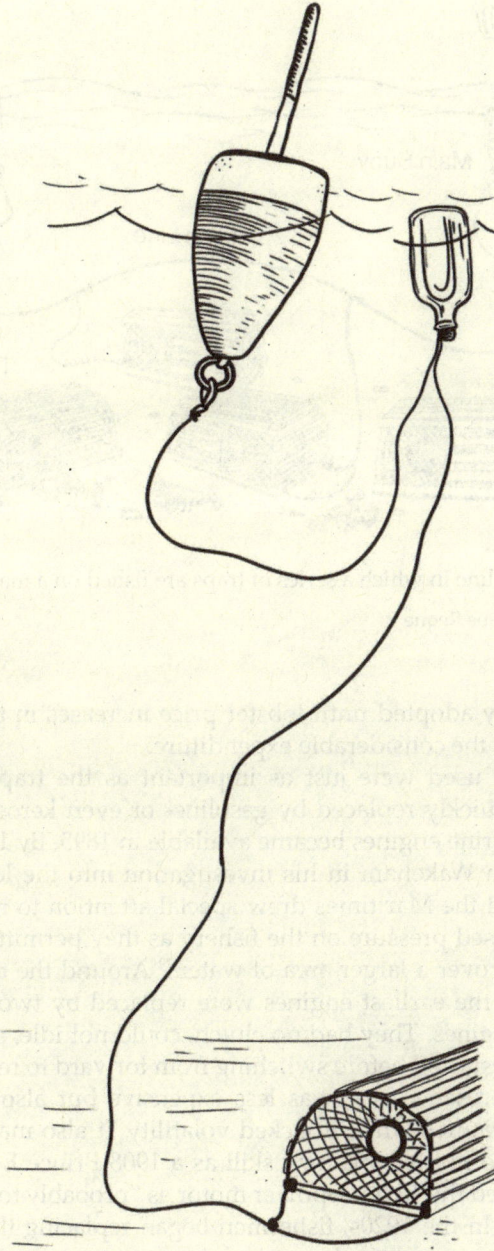

Figure 4. Lobster Trap on Single Line.
Illustration by Marieke deRoos

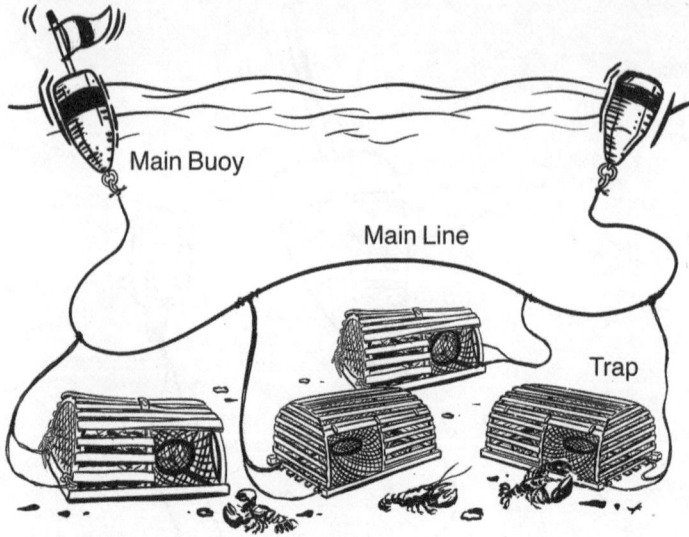

Figure 5. Trawl line in which a series of traps are fished on a main line.

Illustration by Julius Reque

but not widely adopted until lobster price increases in the 1980s and 1990s justified the considerable expenditure.

The vessels used were just as important as the traps. After 1900 dories were quickly replaced by gasoline- or even kerosene-powered boats once marine engines became available in 1895. By 1910 Commissioner William Wakeham in his investigation into the lobster fishery in Quebec and the Maritimes drew special attention to motorboats as placing increased pressure on the fishery as they permitted fishermen to fish deeper over a larger area of water.[29] Around the time of Wakeham's report, the earliest engines were replaced by two-cycle "make and break" engines. They had no clutch, could not idle, and had to be shut off and restarted before switching from forward to reverse.[30] Some boats used kerosene, as it was less expensive but also troublesome at low temperatures when it lacked volatility. It also may have taken a bit of time to master this new skill as a 1908 Prince Edward Island report described the 15 horsepower motor as "probably too powerful ... to manage."[31] In the 1920s, fishermen began replacing these with old motors out of cars and trucks, and engines were adapted gradually to run bilge pumps and mechanically driven trap haulers that facilitated fishing in deeper water.[32] Starting in 1909–10 the Department of Marine and Fisheries collected statistics on gasoline boats, the same as they collected information on other kinds of fishing infrastructure.

Motorboats and fossil fuel were central in the transition to modernity and changed the fishery in two fundamental ways. First, they put pressure on enforcement of regulations. Overseer Abram M. Hatfield from the Yarmouth area wrote to the Deputy Minister noting that all the lobster fishermen used motorboats and as a result the guardians under his supervision rowing in their oar-powered dories were simply unable to enforce the close season.[33] The need for government boats to meet or exceed the technology available to fishermen became a persistent issue. Secondly, for those fishing, motorboats increased expenses and debt obligations. Access to credit tied fishermen to individual canners, buyers, or illegal operators. Gasoline was expensive, especially in the early years of use. In 1907 it was estimated that a 7.5 horsepower engine used a gallon for every seventy-five minutes on the water at 32¢ a gallon.[34] In 1927 one Cape Sable lobster fisherman estimated that he had used $350 in gasoline over three months.[35] The cost of gasoline created leverage for buyers who could use this expense to attract fishermen and create dependence. In the 1930s on Prince Edward Island illegal New Brunswick buyers transported gasoline across the Northumberland Strait to sell to "their" fishermen at the deeply discounted rate of 16½¢ a gallon in exchange for exclusive sales of their lobsters. This compared to the market price of 30¢ (from which a 6¢ fishermen's rebate could be deducted) and made the arrangement worthwhile for fishermen to be tied.[36]

With the increased number of lobster traps in use – made possible by motorboats – there was a corresponding increase in the amount of twine, rope, and bait required to operate the fishery. Twine was used for knitting the trap heads. Hemp or cotton rope preserved with tar was not replaced with long-lasting and low-maintenance nylon rope until the 1960s and 1970s.[37] The amount of rope necessary to link traps to buoys or series of traps in trawl lines was considerable, as after 1899 legal fishing could only be done in at least two fathoms (3.66 m) of water. At one raid at Beach Point, Prince Edward Island, in August 1898, between two and three thousand fathoms of rope used in illegal fishing were brought ashore and stolen by thieves or the original owners during the night.[38] This converts to over five kilometres of rope confiscated in this one raid. In the late summer of 1936 in a small section of the Northumberland Strait, two subdistrict guardians and a Fishery Protective Service vessel seized 86,653 fathoms of rope in the close season. The loss of nearly 160 kilometres of rope in this one small area of coast made no dent in illegal operations.[39]

The over two million lobster traps reported in the 1912–13 Fisheries Branch annual report also required enormous quantities of bait. If fishermen were working on their own, bait, like gasoline, tied fishermen to certain canners or buyers. As the number of traps fished increased, independent fishermen could not supply their own needs, so bait became another cash expense. The quantity of bait required also created a specialized herring fishery for very small fish with its own distinct, but inter-related conservation and ecological consequences.[40] In 1949 a special West Coast herring fishery was opened, and the Department of Fisheries covered shipping of two million pounds of frozen bait across the country by rail when a shortage threatened the East Coast lobster industry.[41]

Those participating in the commercial fishery were almost exclusively from the local white settler community. This created a distinct dynamic compared to West Coast fisheries as until very recently, there were no racialized scapegoats to blame for overfishing lobster.[42] A handful of African Nova Scotian fishermen can be identified in Shelburne County, and the 1881 census lists Richard Hamilton of Port Latour as a Black lobster packer.[43] The Mi'kmaq were among the first commercial lobster fishers in the region. A 1787 Nova Scotia traveller's account described a "great abundance" of lobsters "and the catching them is chiefly confined to the Indians who carry them to markets in their small canoes."[44] Francis Thoma, a fifty-year-old Mi'kmaw man from Esgenoôpetitj/ Burnt Church, New Brunswick, was identified in the 1871 census as a "tinsmith and great hunter."[45] Presumably he employed his valuable tinsmith skills at a lobster cannery. In 1882 the Paqtnkek-Niktuek No. 23 Indian Agent reported that some young men from the reserve had taken summer employment in a lobster factory and in the 1930s, seasonal employment canning lobsters at the Escuminac Point W.S. Loggie and A. & R. Loggie factories provided an important source of employment for the Elsipogtog First Nation at Big Cove.[46]

Although there was an ongoing important subsistence Indigenous lobster fishery, it is striking that the Mi'kmaq who engaged widely in many forms of wage labour were generally shut out of the commercial lobster fishery that took off after 1870. On the Atlantic and Fundy coasts, dispossession pushed Mi'kmaq inland, although they continued to conduct seasonal migration to the shore throughout the nineteenth century. The lack of land meant that boats and gear could not have been stored as, except for small lots in Sheet Harbour and Sambro, Nova Scotia, there were no reserves with saltwater access on the Atlantic and Fundy shores. There is a reason that many men in Kespu'kwitk (Southwest Nova Scotia) specialized in inland guiding.[47]

The situation for the Mi'kmaq on the Gulf of St. Lawrence was no less complicated even though there were some direct, though limited salt-water access on certain reserves.[48] Here the exclusion may have been the result of local merchants being unwilling to advance credit. The Gulf lobster fishery was generally rooted in credit obligations to local merchants, and there were constant complaints about white settler fishermen absconding to escape debt obligations.[49] It was almost impossible to collect debts from settler commercial lobstermen with the aid of the courts, and Indigenous fishermen may have been dismissed by packers and merchants as not worth the risk.

There were at least some Mi'kmaq fishermen who fished commercially. Commercial licences were held at Puksaqte'kne'katik/Pictou Landing First Nation in the 1930s and 1940s onward, and during the 1950s older men found employment in making and selling trap hoops for settler lobster fishermen.[50] At Esgenoôpetitj/Burnt Church in 1950 Mi'kmaq commercial fishermen were identified as "rowing heavy dories to their lobster pots." Anthropologists Wilson D. Wallis and Ruth Sawtell Wallis claimed that in 1953 "quite a few men" in this community used the federal government's Fishermen's Loan program to buy $200 outboard motors for the lobster fishery.[51] The same year the Indian Affairs Branch reported some success with lobster fishing at Lennox Island, Prince Edward Island, where there were a few boats "financed either through a Veterans' Land Act grant or a revolving fund loan."[52] Research needs to be done to know how many of these very small numbers of licences were relinquished during the 1978–83 federal buyback programs. Certainly, this would make sense in light of how attractive cash would have been in a period when lobstering was not making much money.[53] However, coastal territorial dispossession and possible exclusion from credit networks were probably more important to Indigenous marginalization within the commercial fishery. For the period before 1970, I came across no specific mention of "lawless" Mi'kmaq lobster fishermen in any government records. This is significant, as lobster was by far the most frequently policed fishery and Indigenous people were prosecuted for pursuing other fisheries such as oysters, salmon, shad, and trout.[54] The absence of lobster prosecutions, the fishery subject to the most intense surveillance by the state, suggests that – as was the case for many coastal people – lobster was enjoyed by Mi'kmaq discretely outside the commodity stream.

In contrast Acadians had a significant visible presence in the lobster fishery as factory workers, fishermen, and packers. Early Acadian national leaders had a negative view of the fishery. In 1886 Le Courrier des provinces Maritimes published a critique of the lobster fishery as its summer seasonal employment was blamed for Acadian farmers neglecting their land.[55] Later Acadian leaders embraced the fisheries' opportunities as a strategy to preserve the French language and Acadian culture by preventing the outmigration of young people to the United States.[56]

The lobster fishery was also a source of pride for Acadian entrepreneurial success. Olivier Melanson (1853–1926) was probably the first Acadian millionaire when between 1896 and 1905 he owned no less than a dozen factories.[57] Foreshadowing the emergence of Church-led cooperatives in Catholic Acadian communities, Father Pierre Fiset in 1896 bought the Charles Robin Collas Company lobster factory at La Pointe, Cheticamp Island, and operated it until his death in 1909.[58] In the 1910s and 1920s, as large corporations left the region and credit bonds broke with the circulation of ready money, entrepreneurial opportunities opened up for New Brunswick Acadians such as Louis Gauthier of Miscou and John Beaudin of Lamèque.[59] On the water, Acadians were particularly visible among those accused of illegal fishing on Prince Edward Island, and it is impossible to know if this reflected levels of poverty, surveillance, poor political connections, community defiance of a formal law, or some combination of these factors.[60]

Other groups were also identified with the lobster fishery. Bank schooner fishermen on Nova Scotia's South Shore fished lobster seasonally as part of a migratory fishery. In November 1898 the *Boston Globe* reported on the annual departure of the approximately a thousand "provincial fishermen" from Gloucester, Massachusetts, back home to their families and small farms on the South Shore.[61] The end of the bank fishery season coincided with the start of the local legal lobster season, and they were home in time to get their lobster traps prepared for the start of the season on 1 January. When the season ended, they returned to Gloucester in June. At the time of this report, crews on the bank vessels were being replaced quickly by European immigrants, and this seasonal transborder migration largely disappeared with the First World War.[62]

Another group of mobile labour was identified on the Northumberland Strait. A 1903 report stated that many of the fishery's male employees came from Nova Scotia's South and Eastern Shores. They arrived a month before the start of the season and were paid a day wage for repairing gear and making new traps. Once the season opened and they started fishing, they were compensated by the number of lobsters they caught.[63] In the records of New Brunswick's W.S. Loggie Company, it is possible to identify the names of the same families on the payrolls of different factories as they relocated to work in different fishing seasons throughout the spring and summer. In Wallace, Nova Scotia, young, migrant Acadian women worked the canning season in Josh Allen's factory. Local stores would order a "special selection of new women's clothing" to cater to these women who the local museum describes as "more fashion conscious" than the area's permanent residents.[64]

It is possible that the fishery also provided employment opportunities for older male workers. When the Portland Packing Company established a factory in Clark's Harbour, Nova Scotia in the early 1860s, it provided employment for "middle-aged and older men," as most of the young men had left for either the United States or fishing on the bank schooners.[65] This link between lobster fishing and older men was made again in a 1901 Westport, Nova Scotia, protest against the 15 December season start date, as it was a "great wrong to the old men who cannot go out in stormy weather in December, and besides it breaks up the crews of the fishing vessels."[66]

Conclusion

This focus on human lifecycle and the economic ecology of the community, in a curious way, brings us full circle. Just as it is important to understand the humans who fished lobsters commercially and the technologies they adopted; it is essential to comprehend the crustaceans themselves. The conditions under which lobsters thrived and could be safely fished set perimeters for the fishery. The new technologies that were adopted, especially the wooden traps and motorboats, facilitated an intensification of the fishing through more traps, setting them in deeper waters and over larger areas of water. This context is foundational for what follows. As a slow maturing, relatively sedentary cold-water creature, whose safe preservation had initially posed a challenge to its human predators, lobsters were understood to be easily overfished in a local context. As lobster emerged as a valuable new commodity, its possibilities were always constrained by its biology.

Chapter Two

Lobster Becomes a Commodity

The exchange value awarded to lobster had everything to do with time and place. Certainly, early settlers were struck by lobster's abundance. When available, it was used as food, fertilizer, and bait. A mid-nineteenth-century booklet aimed at British immigrants noted that lobsters in Nova Scotia could be purchased for a penny a piece and claimed that a common Halifax sight was newly arrived regiments "enjoying to the full the unwonted luxury" of boiled lobsters.[1] Even then and there, lobsters were already understood to be an extravagance in a wider context, but not locally, and so their worth was contingent on place and circumstance. Access to fresh lobster was restricted to specific places until industrialization transformed something that was seasonally plentiful with very little local cash value into a preserved commodity that could be easily and safely consumed beyond the communities where they were fished and landed. The value of lobster increased, both as a live or canned product, when brought into the broader capitalist marketplace.

From the middle of the nineteenth century into the interwar period, the fishery and the lobster industry were almost synonymous with a canned product. The eclipse of canned lobsters by live ones occurred before the Second World War. Despite it being the same fish and sharing common fishing methods, export markets, and some overlapping fishermen, entrepreneurs, and businesses; the two fisheries or "products" were different in their organization, required infrastructure, working conditions, potential for earnings, and attitudes towards conservation and regulation. The commercial lobster fishery must be understood as having specific chronological periods, distinct geographic contexts, and divergent business, labour, and market structures according to the product – canned or live – being produced and sold. The commodification of lobster was a result of industrialization and advances in transportation. The lobster fishery was a rural site of modernity, not some quaint, traditional fishery. The transformation of a crustacean into a product (or rather products) that was assigned a volatile economic value took place in a specific geographic and temporal context. Both canned and live fisheries were sources of credit and available cash. They depended on waged and unwaged work and provide an important example of how, as Daniel Samson has put it, "capitalist modernity grew within the countryside."[2] Canning, when embedded in a mercantile credit system, left some fishermen perpetually in debt and bound to specific packers who extended credit for gear. But it was also a vital source of "ready money," a result of wages earned in factories, and cash earned from the construction and maintenance of facilities, boat building, fuel cutting, and the transportation of goods to wholesalers.[3]

This chapter will examine three periods in the development of the lobster fishery. Commercial lobster caught in "Canadian" waters was sold to the export market. The first period was roughly 1870 to the First World War and, apart from locations with easy access to urban American markets that imported live lobsters, it was almost exclusively based upon canning. In 1890, canned lobster comprised nearly 88 per cent of the $1.139 million in total lobster exports, and in 1914, while the industry had grown threefold, to nearly $3.7 million in export values, canned lobster still constituted 80 per cent of sales.[4] The market was Europe and the United States, and the industry was dominated by foreign, primarily American producers and British wholesalers. Originally almost all canned lobster was destined for England, but by 1900 the United States was the most important market, representing 41 per cent of sales, followed by Great Britain with 32 per cent and France with 22 per cent.[5] Unlike other fisheries that came under the influence of the Progressive Era belief that sustainability was possible through efficiency and rationality based on modernized industrialized corporate fisheries, lobster did not. The lobster fishery escaped the modernizers' eye as it was labour intensive, poorly capitalized, decentralized, and open to very limited technological innovation.[6]

The second period, between the First and Second World Wars, was connected to changes already underway but cemented by market disruptions associated with the First World War when French and German buyers were lost, and Britain banned the importation of lobster as a luxury good. Here the loss of these traditional European markets, industrial consolidation and reorganization, competition from new products and consumer preferences, transportation and processing improvements, and an assault on packers' monopolies facilitated a rapid decline in the number of canneries and a rise of live market sales. For the first time in 1936, the value of live market lobsters surpassed that of those canned.[7]

These industry changes also faced an expanded regulatory state that adopted a wider range of strategies to bring the industry under its control for the purpose of quality assurance, legal enforcement, and social peace. Finally, after the Second World War until when this study ends in 1970, another form of the fishery emerged characterized by the Canadianization of distribution networks and solidified consumer preferences. Lobster became unique among Canadian fisheries as an inshore pursuit that, at least on the water, often remained outside corporate control. Although lobster persisted as the most important source of income for East Coast fishermen, it largely fell off the federal government's radar, which was focused on industrializing offshore fisheries. Until the rise of new markets in Asia in the twenty-first century, the export market remained almost exclusively the United States. Each of these three periods was associated with specific working conditions for fishermen as direct employees, bound indebted workers, or independent commodity producers.

Lobster was slow to develop as a commodity, as it had little market value until industrial processes solved the obstacle of safe and tasty preservation and overcame the challenge of getting a quickly deteriorating and potentially toxic fish safely to consumers. As noted in the first chapter, lobster could not be easily smoked, dried, or salted, nor could it be kept alive for long periods out of water unless it was cool, wet, and possessed a hard shell. Therefore, lobsters were caught by settlers only for immediate and local consumption until new technological innovations in food preservation altered the spatial and temporal restrictions that had severely limited the development of consumer markets. Yet, in the second half of the nineteenth century, advances in canning technologies, as well as vertical and horizontal transnational corporate structures, gave rise to a new and powerful industrial fishery, but one that continued to be shaped by local geography and the environment.[8] The biological imperative to can lobsters when and where they were caught meant that processing had to be decentralized, in sharp contrast to the canned salmon fishery, which developed at the same time on the West Coast using similar canning technologies.[9] Whereas salmon congregate at specific times in specific places, lobster by nature were scattered throughout their North Atlantic habitat according to sea bottoms, water temperature, ocean currents, and concentrations of phytoplankton. Coastal lobsters migrate relatively short distances inshore and out according to the season to take advantage of their preferred (generally cool) water temperatures. When international prices for canned products were strong and there were lobsters to be caught, this decentralization made sense. Money could be made so long as not too much capital was invested in any facility. The result was the construction of hundreds and hundreds of often very small canneries operating in scattered shoreline locations during potentially very short, state-decreed fishing seasons. Capital investment was dispersed very thinly, with a wide range in scale from large factories through shanty canneries to illegal temporary bootleg operations in the woods.[10] While the production was spread out, the influence of the larger companies was everywhere as they bought and marketed the goods from small independent operators, profited from selling them specialized canning supplies, and even handed off all the risk of the sector through leasing their facilities during volatile market periods.

Lobster fishermen always did something else as the fishery was a seasonal and labour-intensive (rather than capital-intensive) resource extractive industry. Fishing strategies and technology mattered, but the combination of lobster's ecological habitats and life cycle, government regulations, and the consequential structure of the two parallel lobster industries (can and live) resulted in the creation of a decentralized, part-time fishery embedded in local economies already structured around seasonal occupational pluralism.

Lobster as a commodity is characterized with general myths that ignore specific times, places, and contexts. The most frequent general tale in the region is that lobster was the food of the poor and children were taunted for bringing lobster sandwiches to school. It was more complex. When locally consumed, lobster in rural communities was no more expensive than other fish. Elsewhere, it was always a luxury, both when it was canned or available live.[11] By the mid-nineteenth century, it was recognized as a status-bearing luxury item in New York restaurants, and this fashion grew so that by the 1880s, lobster had a close association with the most famous fine dining room in the United States, Delmonico's. This restaurant's invention of Lobster Newburg was almost synonymous with opulence. In the 1900s, until the First World War, New York City "Lobster Palaces" were a popular fad that combined decadent dining with celebrities, money, and a slightly uninhibited environment that eventually expanded to include some middle-class patrons and tourists.[12] But even within fishing communities, framing lobster as the food of the most impoverished needs a context. Certainly, even after the advent of canning, lobster may have been the food of the very poor in moments when glutted markets meant that buyers stopped buying or canneries closed with no warning. Live "tinkers" or "chickens," undersized lobsters that could not be legally imported into the United States, and the many lobsters caught outside of the legal season could have little exchange value as they were outside most established commodity streams. The infamous school lobster sandwiches had a context as they were made from lobsters that in that time and place had no commercial value as they were extra-legal, or the local markets had failed. Moreover, lobster fishermen who fished as independent commodity producers engaged in the live lobster market made more money, received cash immediately, and got to sleep in their own beds, making lobster an attractive alternative to other fish. If there was ready money available for lobsters, it was not consumed within the household.

Canning Lobster

From the mid-nineteenth century until the mid-1930s, most lobster caught in Canada was canned. Advances in food preservation from France, Scotland, and England in the early nineteenth century eventually were applied to lobsters, and it was found that lobster could be prevented from spoiling through heat and the creation of a vacuum seal. Those working in the area had no idea why this worked. Until the advent of bacteriology in the 1880s, the focus was almost exclusively on the chemical composition – in particular, the amount of acid present – in each food, not on sanitary conditions.[13]

The process that evolved for packing lobsters was extraordinarily labour intensive, as the extraction of meat from shells could not be mechanized. Thousands of "skilled" men and "unskilled" women and children found brief seasonal employment in the packing industry. At first, the emphasis was on the can itself, and the tinsmith was the most valued employee. Individually cut pieces of tin were bent with physical force over wooden patterns and then soldered together. Thinner-walled cans with a rim around the top came into use around 1865, which meant that cans could be opened with a can opener rather than the traditional hammer and chisel. These cans were then filled with cold, cooked lobster claws and tails, capped except for a small hole for venting, and then boiled. The final step was to place a drop of solder over the vent hole.[14]

Individual experimentation with the process created improvements (and problems) through constant innovation.[15] Calcium chloride was introduced in some operations by the late 1860s, as it was found to reduce cooking times from six hours to thirty minutes.[16] Although a locking-seam can that did not require soldering was introduced in 1888, it was not widely adopted until the early twentieth century as its cost could not be justified by small operations.[17] As late as 1909 the value of a skilled male "sealer" was essential if any lobster was to be processed.[18] Safety and consistency remained constant problems, although a steam retort (which acted like a pressure cooker) was patented in 1874. The retort had the capacity to accelerate the cooking process, and the reduced cooking times still killed bacteria but improved taste (and texture) a great deal. Wealthier American canners, as well as certain large Canadian and British companies, gradually introduced these machines. In 1922 a good steam retort cost between $400 and $500, too much for many packers.[19] In the 1920s concerted efforts by larger firms lobbied for the mandatory use of this more sanitary equipment and the government responded by expanding standardized inspections.[20] By 1934 the number of factories without a steam retort had fallen to 28 per cent from 54 per cent in 1927.[21] Despite these innovations, safety and consistency in quality remained a constant problem.

One of the first known examples of lobster canning (and salmon canning) took place in the early 1840s at Portage Island, at the mouth of the Miramichi River in New Brunswick.[22] We know that canning took place on Prince Edward Island in 1858, yet it did not take off quickly there, as the 1874 report of the federal Fisheries Branch stated that only 1,443 cans of lobster were processed during the Island's first year in Confederation. In comparison, that same year New Brunswick reported producing over two million one-pound cans.[23] Processing lobsters began early on Nova Scotia's South Shore with American and local entrepreneurs teaming up with skilled tinsmiths in Port Mouton by 1851 and nearby Liverpool.[24] In 1841, Charles Mitchell, a Scottish immigrant and tinsmith, was canning on Nova Scotia's South Shore but appeared a year later in Eastport, Maine, taking his secret processing knowledge with him.[25] Secrecy and specialized expertise was essential to maintaining control over the sector. At the pioneering firm of Treat, Noble and Holliday, of Eastport, Maine, in the 1840s, no outsiders were ever permitted to enter the "bathing rooms" in case competitors gained trade secrets. A policy of restricted entry remained in place twenty years later when the first cannery opened at Clark's Harbour, Nova Scotia. Here, one room in the factory was kept locked and reportedly no one except the manager was permitted to enter.[26]

While Mitchell's expertise went south, American capital and skill soon moved in the opposite direction. There were never that many canneries in Maine, as the decline in lobster stocks there prompted American entrepreneurs to open the Canadian fishery. Maine boasted a peak of twenty-three canneries in 1880, but these had all closed by 1895, with the state's introduction (and enforcement) of a conservation-oriented size limit.[27] Depleted stocks led Maine and Massachusetts canners to move into the Maritimes, the Gaspésie, and eventually the west coast of Newfoundland, bringing their employees, expertise, capital, and distribution networks with them. The same year that the Canadian government struck its first commission on the lobster fishery, the US Commission of Fish and Fisheries published a massive report that included the "British Provinces" in its discussion of the American lobster canning industry. While five Portland- or Boston-based companies owned twenty-three canneries in Maine, they operated more than forty north of the border by the mid-1880s. The 1887 report explained that these were established after 1870, when foreign demand exceeded what the Maine fishery was able to meet, and that these products were entirely shipped directly overseas to avoid an 18 per cent US import tariff. That tariff on canned lobster was greatly reduced in 1890 and disappeared completely in 1913, expanding markets among American consumers.[28]

The near invisibility of lobster canning in the early days of the Maritimes is evident in an article that appeared in the *Halifax Citizen* in 1863 describing the operations of the American lobster packing firm of Hamblen, Baker and Company, of Sambro, Nova Scotia, which had been in operation since 1857. In 1862 it processed over 200,000 hermetically sealed tins of fish, with no particular emphasis on lobster described by the newspaper. The firm was said to employ about 150 men and boys and twenty-five women from April to December, although "little seems to be known among our citizens." This public indifference was shaped by the novelty of the product and scant technical expertise, perhaps also explaining why so few local entrepreneurs were part of the first wave of packers.[29]

This unfamiliarity would not persist for long, as the industry experienced explosive growth in the 1870s. By the early 1880s, lobster was the most valuable fishery, after groundfish, on Canada's Atlantic coast. At the end of the century lobster had surpassed important species such as cod and haddock in export value. Most of the lobster canned originated in the Gulf of St. Lawrence. By 1900 the Prince Edward Island lobster industry – exclusively connected to canning – was the second-most-important economic activity in the province, behind only farming.[30]

The incredible potential for profits in the early years of lobster packing was a powerful incentive to encourage new canneries along waters that had not previously been fished. In 1872 there were forty-four lobster canneries in New Brunswick, Nova Scotia, and the Gaspésie, almost all of them operated by just five New England firms.[31] The two most important companies, the Portland Packing Company and Burnham and Morrill, both originated in Portland, Maine. From the northeastern coast of the United States, ambitious canneries moved northward to the Atlantic coasts of Nova Scotia and New Brunswick,

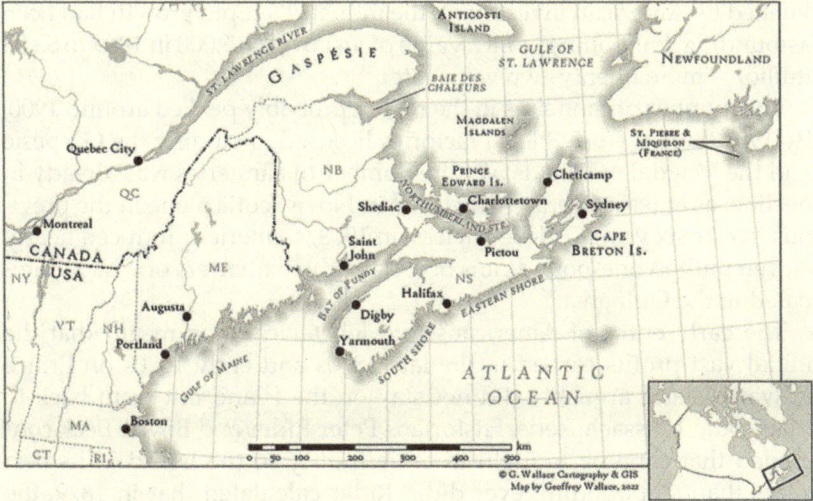

Figure 6. Map of the Gulf of Maine and the Gulf of St. Lawrence linked Massachusetts to Quebec's North Shore, the Maritimes and Western Newfoundland as a lobster region.

© G. Wallace Cartography & GIS

and once those areas were in full production, on into the Gulf of St. Lawrence. They spread from Nova Scotia's South Shore north to the Eastern Shore (Guysborough), then to Cape Breton. No canneries operated on the western Gulf coast of Cape Breton until around 1878, when H.S. Andrews of Essex, Massachusetts, who had had operations on the Eastern Shore, started packing there. He eventually sold out to Burnham and Morrill.[32]

J.H. Meacham's *Illustrated Historical Atlas of Prince Edward Island*, published in 1880, appears to have inadvertently captured this moment of rapid transition on the Island. Its prospectus suggests that the maps are based on surveys conducted in 1877 or 1878, and around thirty-five lobster factories can be identified on the Island. Almost all the factories have specific local names attached to their establishments, and only I.B. Hamblen and Sons, at Wood Island on the eastern end, can be linked to American operations. Yet Americans were there. By 1885 the Portland Packing Company, which was consolidating its operations in the Gulf of St. Lawrence and the Northumberland Strait had transferred W.F. Tidmarsh to Prince Edward Island to establish its operations there.

Fuelled by American investment, the industry's rapid growth had been astounding, ballooning from a value of just over $15,000 in 1869 to $2.25 million a mere twenty-two years later.[33]

The number of canneries in the region probably peaked around 1900. By 1909 there were 723 legal factories licensed, including the Gaspésie and the Magdalen Islands; yet the number of canneries was already in decline, as at least twenty had closed in Nova Scotia alone in the previous five or six years.[34] Nevertheless, in 1908, canneries produced nearly eleven million one-pound cans of lobster, three quarters of which originated in the Gulf area.[35]

The early entry of Americans into lobster canning meant that the initial vast profits earned in the late 1870s and early 1880s on Prince Edward Island arguably did not stay on the Island but went back to Maine and Massachusetts. Historians Peter Rider and Boyde Beck concluded that canning brought in more money to the Island in its best years than shipbuilding ever did.[36] Rider calculated that in 1879, the thirty-five canneries in operation on the Island earned about $13,285 per facility. With a proliferation of operations in the 1880s and declining catches, profit margins fell so that by 1892, the 212 canneries in operation earned an average of only $1,863 per factory.[37] As the first cohort of operators, Americans made the most money. Nevertheless, lobster remained essential for the survival of coastal communities as shipbuilding disappeared and alternatives such as the Gulf mackerel fishery failed.[38]

Profits declined as catches fell. Even though the government's fisheries bureaucrats were obsessed with counting, their statistics are notoriously unreliable before 1917, warped by such factors as variable time frames (end-of-year changing from June to December), inconsistent units of measurement, under- or over-reporting by fishermen and canners, and the imagination, incompetence, and political biases of bureaucrats, inspectors, and scientists.[39] The Canadian Department of Trade and Commerce's record of exports of one-pound cans was also an inconsistent run of serial data. While the unit of a "case" remained constant at being comprised of forty-eight one-pound cans, what was meant by a one-pound can changed over time. Until around 1890, a one-pound tin was sixteen ounces of meat, but this fell so that by the 1930s, a pound can might contain ten ounces of dry lobster meat and the rest liquid brine.[40] This needs to be understood when one reads that Canada exported nearly two million one-pound cans in 1873, and in a little more than ten years had increased this number sevenfold. The number of cans decreased in the late 1880s, benefited from a boom in the mid-1890s, and settled down to relatively stable and slightly

Lobster Landings in Metric Tons, by Province, 1893-1968

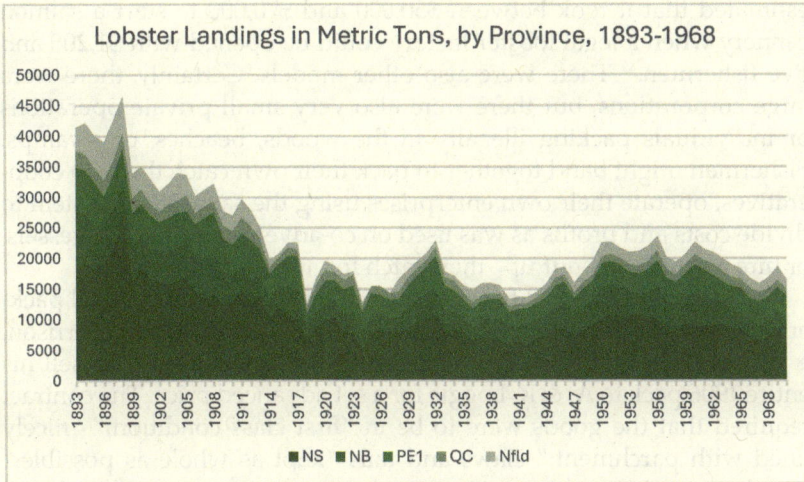

Figure 7. In 1992 a researcher working for the Department of Fisheries attempted to reconstruct lobster landing from 1893. This graph does not capture the boom of the 1870s and 1880s but breaks down landing by provinces for 1893–1970. A.M. Williamson, *Historical Lobster Landings for Atlantic Canada, 1892–1989* Canadian Manuscript Report of Fisheries and Aquatic Sciences 2164, 1992, 19–20.

declining numbers until the First World War. In actuality, after 1883, notwithstanding an intensification of fishing efforts through greater numbers of more efficient traps and, after 1900, the gradual adoption of the motorboat, catches fell, reaching a low point in 1918. Apart from two modest peaks in the early 1930s and early 1960s, landings did not reach 1880s levels for a hundred years, until the 1980s.[41]

Canners

The variety of firms participating in packing during the first fifty years ranged from large transnational American and mid-sized Canadian operations to miniscule, local, legal, and illegal operations. Operating a small factory was within the means of at least some fishermen and certain small-scale coastal village storekeepers. The geographic dispersal necessary for processing meant that the industry was completely different from the development of salmon canning in British Columbia. Salmon was characterized by a concentration of ownership and large amounts of capital necessary to enter the business. In 1909 it was estimated that it took between $50,000 and $70,000 to start a salmon cannery when a legal lobster factory could be opened with $1,200 and five fishermen.[42] There were also other models. Certainly, there were large corporations, but there were also very small private operations or individuals packing illegally in the woods, beaches, or swamps. Fishermen might band together to pack their own catch through cooperatives, operate their own enterprises using the same share system to divide costs and profits as was used on co-adventuring fishing vessels, or hire a canner to "put up" their catch for an established fee.[43]

Credit obligations tied fishermen to canners and in turn small packers to larger concerns. Small village merchants such as Alfred Morrison, a lobster packer at Church Point, New Brunswick, agreed to sell his entire 1906 pack to A. & R. Loggie for a set advanced price. The contract required that the goods were to be in "first class condition," "nicely lined with parchment," claws and tails "kept as whole as possible," and that Loggie would provide empty cans, cases, covers, and solder.[44] Occasionally these small operators canned other items. Some packers in Prince Edward Island tried meat in the 1880s, while the Poole Family of Arcadia, Yarmouth County, canned blueberries.[45] Large packers with significant amounts of capital invested in sanitary facilities regularly complained that smaller, independent operations could not ensure a quality product. In their criticisms, they made no distinction between bootleg or illegal canners and small independent operations and in fact participated in a system where they supplied both these streams of operations with material such as cans and purchased their finished goods to resell under their own brand labels.[46]

Portland, Maine, home of the Portland Packing Company and Burnham and Morrill, played an especially important role in the industry. Together with another Portland company, John Winslow Jones, a packer that controlled the Magdalen Islands during the early years, the Portland-based businesses transferred an expertise developed to preserve vegetables to their Canadian lobster enterprises. Their Canadian facilities rarely expanded to any product beyond lobster, but their diversified corporate structure meant that they could easily weather the volatile markets often associated with canned shellfish.[47] Other Maine companies followed, such as William K. Lewis, J. Henry Forhan, D.W. Hoegg, and Freeman Kimball, along with Massachusetts-based organizations such as Shedd and Knox.[48] But the Portland Packing Company and Burnham and Morrill remained the dominant firms. In combination with two other US companies, H.C. Baxter Brothers and the J.H. Myrick Company, they controlled in western Prince Edward Island, according to one packer, as much as 40 per cent of the entire sector by

1909.[49] Their influence permitted them to collude in annually setting prices or wages paid to fishermen, as well as establishing the wholesale prices that smaller packing operations received for their finished goods and paid for supplies. Politicians faced regular concerns about the "lobster combine" or "cartel," and these appear to have been justified, as there are suggestive examples of competitors cooperating on setting prices to be paid to fishermen.[50] In 1905 former Nova Scotia premier and now Liberal Member of Parliament, W.S. Fielding reported to the Minister of Marine and Fisheries that there was a "combine" being formed in the United States, with the cooperation of Canadian packers, for the purpose of "controlling the lobster industry in the Maritime Provinces."[51] The matter was raised in the House of Commons in January 1908 when John Lefurgey, a Conservative, accused foreign corporations of acquiring lobster factories on Prince Edward Island for the purpose of creating a monopoly. The Portland Packing Company immediately went on the defensive, acknowledging that it had recently purchased four canneries in the member's riding but that it did not own as many factories as it had fifteen years earlier. In January 1913, however, W.F. Tidmarsh of the Portland Packing Company wrote privately to J.H. Myrick suggesting that it might be "a good idea for us to get together" and agree on a maximum price to be paid for lobster. The following year Tidmarsh denied the existence of an organized combine in his testimony before the 1914 British Royal Commission on the Dominions, but he had clearly participated in price setting.[52]

Outside the industry, concern persisted. In 1913 Fisheries Inspector Ward Fisher reported to the Superintendent of Fisheries W.A. Found that a recent Yarmouth meeting of all the large Nova Scotia firms had created a "holding company" aimed at controlling production and regulating the local can and twine markets. The Fisheries Inspector blamed the banks, who had refused to extend further credit to the packers unless they cooperated to stabilize the industry.[53] This desire for solidarity was behind Roberts, Simpson and Company's open campaign for "amalgamation" or consolidation in the canning sector. Its management argued that a monopoly was in the best interests of consumers, fresh lobster shippers, packers, fishermen, and government. According to this company's logic, with a monopoly, consumers would be guaranteed a quality product, those in the live trade would be freed from "the mercy of foreigners" to set prices, and packers would benefit from the consolidation of facilities and efficiency. Moreover, the argument continued, fishermen would get predictable prices and buyers would eliminate the influence of self-interested "agitators." Finally, government through a reduced number of packing licences would achieve its conservation goals with the cooperation of canners.[54] Not surprisingly, the real or rumoured emergence of combines generated a range of political responses from fishermen, politicians, and civil servants. The most important was the policy to grant fishermen's associations or cooperatives, their own canning licences so that they might work outside any monopoly.[55]

The canned lobster market was characterized by extreme volatility. Surplus production glutted markets and resulted in dramatic price drops. Conversely, poor fishing seasons and subsequent price increases generated great profits for canners. Specific years, such as the period 1889 to 1892, witnessed a great many bankruptcies, and a self-interested representative of Roberts, Simpson and Company estimated in 1907 that in any year a single cannery might lose $2,000 to $3,000.[56] By the 1880s, after underwriting and managing the first generation of canneries, all large canners were likely to have acquired an additional revenue stream. For instance, they sold supplies (such as premade cans, twine, and eventually gasoline) to small local operators, and in turn purchased their canned lobster according to contracts made at the start of a season, labelling them as their own and exporting them to European and, eventually, American markets. This placed most of the risks on the small operators and, in lean years, gave what profits existed to the larger companies. Small local packers, often with precarious profit margins, were frequently in debt to larger concerns, and some referenced this as the reason they turned to illegal operations.[57]

The American companies also had diversified income streams. In an unstable market, being able to spread out your operations with other products and sources of income was clearly a corporate advantage. Large American canners were uniquely privileged and protected, for while they had a great deal of capital invested in the region's largest facilities and used the most up-to-date technologies, packing lobster was only a small part of their overall business operations. The Portland Packing Company, Burnham and Morrill, J. Winslow Jones, and Forhan all had assorted operations in Maine, with the most important branches of their companies dealing with the preservation of sweet corn.[58]

Roberts, Simpson and Company, founded at Halifax in 1880 by Frank Roberts, a native of New Brunswick and Liverpool, England-based William Muirhead Simpson, adopted a different strategy. It became the largest wholesaler of canned lobster in the world with branches in Yarmouth, Charlottetown, Shediac, St. John's, and Vancouver and in 1907 moved its head office to Liverpool, England.[59] Less visible than its American counterparts, Roberts, Simpson and Company operated under different names and opened and shuttered its own canning operations according to market conditions. It was the first of the large companies to rid itself of actual processing factories – making money through leasing, supplies, and wholesale. Most large packing concerns came to realize that through leasing their facilities to independent local operators, who were tied to the owners for the sale of their product, they were able to rid themselves of much of the risk with quality, prices, and oversupply.

American companies, such as Burnham and Morrill, began to adopt this model before the First World War. Earlier American operators also realized that they could make more money if they concentrated in specific geographic areas and gained control of the local market. By the 1890s, Burnham and Morrill focused almost exclusively on the Northumberland Strait and western Cape Breton.[60] The Portland Packing Company, which at one point operated eighteen branches in the Maritimes and Newfoundland, ended up clustering operations on Prince Edward Island. This same scramble for geographic concentration had led John Winslow Jones to the Magdalen Islands and the west coast of Newfoundland in the 1880s, while D.W. Hoegg operated on the Baie des Chaleurs.[61] The concentration of specific companies in specific geographic locations could have devastating results. The bankruptcies of the William Leslie and Company and Eastern Canada Fisheries in the 1920s was a hard blow to the Magdalen Islands, as these companies had almost complete control of the Islands and little cash circulated as households were tied to these merchants.[62]

The scale of most American operations meant that they had invested substantial sums of capital. The 1887 report published by the US Commission of Fish and Fisheries estimated that Americans had invested the remarkable sum of $12,500 per cannery.[63] In contrast, that same year the *Mercantile Agency Reference Book* listed eleven non-American packers on Prince Edward Island, four of which had credit value of less than $1,000. Scores of short-term packers must have been too small for the credit agency even to notice. The 1891 Canadian census reported that the fixed capital in "fishing canning" establishments averaged from $691.92 per cannery in New Brunswick to $1,216.15 per cannery in Nova Scotia, with Prince Edward Island averaging $1,099.14 per factory in land, buildings, and tools.[64] An 1892 report was even more disparaging. It quoted a Prince Edward Island fisheries inspector as stating that most factories were run "absolutely without capital": "a rough shanty, a lobster dory (worth about $8) for every 100 or 150 traps; some laths to make traps, and [a] few logs of firewood are all the implements required for a start." These miniscule operations were usually operated on advanced credit by a local merchant, the merchant in turn tied to a larger American, Canadian, or British exporter.[65] The extreme differences in capital invested, and the corresponding uneven condition of equipment and facilities, created a range in product quality. In particular, American canners and federal fisheries officials held these small-time packers responsible for Prince Edward Island's reputation in the British market for poor-quality lobster.[66]

There were Canadians involved in packing from almost the beginning with midscale operations. W.S. Loggie, O'Leary, Matthew and McLean, and J.W. Windsor began between the 1870s and 1890s and bridged the interwar period. At one point, W.S. Loggie operated forty-five factories between Miramichi and Lamèque in New Brunswick.[67] The most eccentric was R.B. Noble of Richibucto, New Brunswick. The labels of his "Crown Brand" canned lobster were an expression of his Christian evangelicalism featuring a boiled lobster and Bible verses. A reporter writing in the 1950s noted that "there is something disconcerting about an image of a bright red cooked lobster beneath the words "Christ in You the Hope and Glory."[68]

Across the Gulf of St. Lawrence in Newfoundland the structure of the lobster industry looked different again. Until after the Second World War, when Newfoundland joined Canada, it was, apart from a few west coast communities after the late 1920s, an exclusively canned commodity. After large American and Canadian companies withdrew from Newfoundland in the late nineteenth century, most canned lobster was preserved at the household level, following a model similar to that used for processing cod. With a meagre $50 investment in capital goods for canning supplies, packing at the household level kept lobster production at a very small scale and produced uneven quality.[69] It also meant that there was not an organized business lobby to oppose the Newfoundland government's introduction of a three-year moratorium on the lobster fishery in 1925, 1926, and 1927. To move away from domestic production and encourage fishermen's cooperatives, in 1930 the Newfoundland government decided that packers who had not processed fifteen cases in 1929 would not be given a new licence.[70]

In the Maritimes and the Gaspésie, households might participate in the knitting of trap net or whittling wooden lobster plugs to immobilize claws but legal production, at least, took place outside the household. Like the home, work there was strictly divided by gender. Men worked on the water, moved the catch from the wharf, and took care of the fires and boiling rooms.[71] Even though the season did not last long, cannery work was an important source of cash for rural women and often the only alternative to domestic service. Women

LOBSTER CANNING.

Figure 8. Robert Harris, "Lobster Canning" c. 1882 in *Picturesque Canada: The Country As It Was and Is*, George Munro Grant, editor (Toronto: Belden, c. 1882), 859. This clean, romanticized image was the Canoe Cove, PEI cannery operated in 1880–1 by the artist's father, William Critchlow Harris Sr. Courtesy, Special Collections, Killam Library, Dalhousie University.

participated in the lobster fisheries in both unpaid and paid capacities. At home, they helped knit nets for traps or might participate in the processing of illegal bootleg canning operations. In factories women and girls worked in specific positions as a rigid sexual division of labour was maintained. Women were involved in cooking and packing, tasks strongly associated with women's unpaid domestic labour in the home. Daughters of fishermen found their first employment in the factories, and their ongoing employment might be part of an understanding of to whom their fathers' lobsters would be sold. Into the twentieth century, daughters of the best fishermen were usually the last to be laid off.[72] Sketches, paintings, and photos of factory workers highlighted women, and employment in lobster canneries was identified as a form of industrial work of women in a 1900 international publication.

Women working in lobster factories, like women in factories elsewhere, were undervalued and poorly paid. According to the 1889 *Report of the Royal Commission of the Relations of Labor and Capital*, men working in canning factories received $25–40 a month with board and lodging, and women received $12 a month without board or $8 with board and lodging.[73] There was one important exception. If the most valuable employee at the beginning of the industry was the skilled tinsmith who ensured a quality product, with time, this hierarchy changed and the most important employee in a cannery was said to be the woman who cooked for the crew. As factories fed their employees during the short intense season, good, plentiful food was essential for the retention of contented workers.[74]

On the water, lobster fishermen worked under a variety of labour regimes. In the Gulf region, some were wage employees, fishing for a specific canner. Others were contracted by the season, either with or without gear and boat, and bound to sell their catches exclusively to one packer. The *Moniteur Acadien* reported in 1886 a variation of this where the local merchants advanced credit on supplies and held the bill of sale and the fishermen's gear as security.[75] The rates of return varied greatly according to region. One historian has estimated that, in 1886, lobster fishermen in Prince Edward Island received 25¢ per hundred pounds of lobster, while the same amount fetched fishermen selling live lobsters in western Nova Scotia $10–20, depending on the market and time of year.[76] On the South Shore of Nova Scotia and the Bay of Fundy, fishermen usually fished as independent commodity producers, owning their own boats and gear, and were free to sell their catch to the highest bidder. For these independent fishermen, the appeal of lobster fishing was connected to the increasing difficulty fishermen had in making money off other fish. Cod fish had become "so unremunerative" that the *Yarmouth Herald* in 1881 claimed that almost all the fishermen in western Nova Scotia had invested in lobster pots.[77] Lobster earned money. A 1909 federal government investigation made the claim that a "considerable" number of New Brunswick and Prince Edward Island lobster fishermen were "not bona fide fishermen" but poor farmers who "find it difficult to get ready cash." Lobster fishing was the first employment in the spring, and payments were made in cash if there were no debts to clear.[78]

Live Lobster

Live lobsters were a distinct commodity, with a different geography, labour, and distribution structure. While the same fish, live lobsters were worth also a great deal more money to fishermen. In areas where, and in times when, they overlapped with canned lobster, fishermen – who might fish both – did not share the same interests with packers. Live market lobsters had to be a specific size in order to be legally imported into Massachusetts or New York. This created two streams where catches might be sold: market size for international export or smaller "canners" to local canneries. Market-sized lobsters received a much higher price and so from the moment this option emerged, there was an economic incentive to support size limits and encourage the growth of a stock of larger lobsters. The parallel market for live lobsters meant that fishermen who had this option were not direct employees of canners and usually had some degree of autonomy to whom their catch might be sold. This independence, of course, could have limits and might be shaped by credit obligations, lack of feasible transportation infrastructure, or competition among local buyers. Railway access to Yarmouth, an important transhipment centre, permitted fishermen to ship live lobsters themselves directly to American agents or customers.

Before the First World War, the live market represented a relatively small proportion of the lobsters caught in Canadian waters but was an important source of revenue for those who could participate in it. From the early nineteenth century onward, there was an established market for live lobsters in American coastal cities. This demand resulted in overfishing in waters off Southern Maine, Massachusetts, and New Hampshire. Consequently, the catchment area expanded northward by the 1880s, as lobsters caught in the cooler fall, winter, and early spring fishing seasons could survive – often wrapped in seaweed or carried in the wells of smack boats (vessels with built-in tanks filled with seawater) – the long voyage from the Bay of Fundy or the South Shore of Nova Scotia to New York or Boston. By 1882 the Arcadia Lobster Company out of Yarmouth had built its own smack, and in 1887 the Yarmouth Lobster Company delivered live lobsters from its home to Boston in just seventeen hours.[79] An American fisheries investigator noted in 1895 that fishermen on the Bay of Fundy and Nova Scotia's South Shore were shipping lobsters themselves directly to Boston or neighbouring states. These consignments were made up of wooden boxes of 140 pounds of lobster with 220 pounds of ice. Fishermen paid the cost of freight and the commission and bore the risk of the lobsters arriving alive, but they could also make more money than from any other kind of fishing. They also learned to benefit from currency differences between the US and Canadian dollar.[80] The extent of this practice is suggested by 1899 newspaper advertisements placed by the New England Lobster Co. of Boston and Newport, Rhode Island, in the

Yarmouth Herald. According to state law, and the different minimum import size, fishermen were told to ship their lobsters under ten and a half inches to Rhode Island, while larger lobsters were to go to Massachusetts.[81] After the completion of the Halifax and Southwestern Railway in 1906, fishermen along the Atlantic coast consigned their catch via Yarmouth to agents in Boston.[82] By 1911, the Canadian fisheries' bureaucracy, if not its political masters, realized this was the future. W.A. Found, Superintendent of Fisheries from 1911 until 1938 and someone very familiar with the lobster industry, wrote in a department memo that "the day will surely come when canneries will be a thing of the past all over." He noted the current demand for live lobsters and predicted a future when cold storage would permit cooked lobsters to be shipped in the shell.[83]

Fishermen in the live market sector, who did not ship themselves, were usually able to sell their lobsters in a competitive market. American smacks began calling on Nova Scotia's South Shore in the early 1880s, disrupting the local canners' control of the market. When the smack *Pride of the Port* arrived in Clark's Harbour in 1881, the vessel's captain offered fishermen 5¢ for each large lobster compared to the $1.60 per hundred count they received from the cannery. Moses H. Nickerson, who was employed as the factory bookkeeper at the Portland Packing Company, was instructed to inform fishermen that if they sold their large lobsters to the American smack, the cannery would not buy those under ten and a half inches. The fishermen were not daunted, and after two days of no deliveries to the cannery, the Portland Packing Company backed down. Two years later, the factory closed its operations at Clark's Harbour, where live market lobsters had become the most important catch very early on.[84]

But even independent commodity producers could be tied to buyers through formal and informal ties. In 1905 the *Liverpool Advance* reported that fishermen in Little Harbour and Port Joli had been forced to sign a "queer document" produced by the Canadian Atlantic Canning Company of Lockeport. This contract dictated that until the American-owned smacks arrived in the spring, fishermen would exclusively sell their medium-sized lobster to this company for live export. The penalty for breaking this arrangement was $500, as Canadian Atlantic Canning had chartered its own vessel to collect lobsters early in the season.[85] Lobster buyers could be independent agents or employed by specific companies. They also leveraged relationships with fishermen that went beyond money. Wilfred Poirier, who worked as a lobster buyer for a Boston company in the 1930s on the Eastern Shore and Isle Madame provides one example. Poirier was hired for his bilingualism and as a practising Roman Catholic, the local priest in West Arichat instructed parishioners to ship through him.[86]

Big Changes: The Interwar Period

Superintendent of Fisheries W.A. Found was correct in his prediction about market trends, but he could not have foreseen how rapidly the lobster industry would be transformed by improvements in truck transportation, road infrastructure, and refrigeration technology. The First World War severely disrupted canned lobster sales to Europe, and the loss of these traditional markets, combined with industrial consolidation, rising costs from new government quality standards, and shifting consumer preferences, fundamentally altered the cannery landscape. In response, some fishermen formed cooperatives to can lobsters themselves, challenging the dominance of packers who had previously held monopolies in certain areas. The decline in the number of factories was dramatic. By 1928, for example, there were less than half as many canneries on Prince Edward Island as there had been in 1900.[87] While the export value of canned lobster remained strong throughout the 1920s, increased regulation aimed at improving quality may have put small operations, which were already operating at the narrowest of profit margins, out of business. Larger economic and cultural factors were also at work. Consumer preferences moved towards the cheaper (and probably more consistently higher quality) tinned Japanese crabmeat, and eventually for fresh or frozen prepared lobster meat so canned lobster supply often exceeded consumer demand.[88] Meanwhile, the market for live lobsters grew so quickly that by the late 1930s, its value surpassed that of canned products. Live lobster was sold almost exclusively to the United States, where Canadian lobsters were exempted from protective fish import tariffs, perhaps reflecting the political weakness of Maine fishermen versus the influence of Boston-based urban fish wholesalers.

Traditional lobster packers were not always able to sell their product. In the aftermath of the First World War and the following recession, 25,000 cases of the 40,000 cases produced in Prince Edward Island in 1920 went unsold and were carried over as surplus to the 1921 season.[89] Just as Americans had been quick to abandon their Newfoundland operations between 1880 and 1885 when that lobster fishery experienced a downturn, they adopted the same strategy on the western side of the Gulf. In the 1920s American canners left the region, initially leasing operations to local entrepreneurs. In 1916 Burnham and Morrill owned eleven factories on the Northumberland Strait, but by the end of the 1920s Canadian canners such as William Broidy's Maritime Packers

Limited, Fred Magee, and Émile Paturel dominated the area. The Portland Packing Company remained an important presence on Prince Edward Island in the 1920s, but the Charlottetown *Guardian* frequently referred to the area manager, W.F. Tidmarsh, as a "lobster buyer," not as someone supervising numerous packing concerns.[90]

At least part of the explanation for the loss of market was associated with ongoing issues around selling a consistent quality product. The problem of why some lobster "blackened" in cans was the subject of a 1919 federal inquiry by Professor A.P. Knight who concluded that past regulations had not gone far enough or been enforced to ensure high quality.[91] The large companies attempted to eliminate small independent operators and successfully lobbied the Department of Marine and Fisheries to bring shellfish under the federal *Meat and Canned Foods Act*.[92]

Canners who stayed in business responded through innovation and reorganizing the structure of their industry. By the late 1920s, packers were experimenting with preparing frozen and vacuum-sealed fresh chilled meat for short-term consumption, making a product much more appealing to North American consumers.[93] The elimination of the extended heat processing improved the texture and taste. In the late 1920s processers such as Austin E. Nickerson of Yarmouth experimented with brine freezing lobster in sealed, vacuumed-packed containers.[94] Shediac-based Paturel and Pictou's Maritime Packers both claimed that they created a market for perishable but hermetically sealed tins of fresh cooked lobster meat assembled without heat processing.[95]

Refrigerated railway cars started shipping live lobster and chilled meat from Pointe-du-Chêne near Shediac in 1928.[96] This opened the Northumberland Strait area and eventually Prince Edward Island to products other than canned lobster.[97] As Depression-era American hotel restaurants shifted toward serving smaller, more affordable lobsters, Canadian suppliers capitalized on a booming market of lobsters under ten and a half inches – sizes that could be legally fished in Canada, but were prohibited in Maine.[98]

The most important transformation for the region was the expansion of trucking, made possible by asphalt and greatly improved roads. This not only promoted the consolidation of some canning factories but more importantly, facilitated the ease of shipping live lobsters. In 1931, local priest Father LeBlanc helped the Fishermen's Association of Buctouche, New Brunswick, ship live lobsters by trucks to Boston.[99] Prince Edward Island, as an island, had more of a challenge, but the launch of the new rail and car ferry, SS *Charlottetown*, in 1931 (to replace the railcar-only SS *Prince Edward Island*) and its subsequent enlargement in 1938 meant that a limited number of vehicles could be accommodated. The new ferry did not solve truck transportation problems (this continued to be an issue until the 1950s) and Prince Edward Island remained disproportionately tied to canned lobster. The *Canadian Fisherman* in 1933 reported that while 23 per cent of the total Canadian catch was sold live, market live lobsters comprised only 6 per cent of Prince Edward Island's catch.[100]

Unless a local fishermen's cooperative was in place, the gradual expansion of the live sector into the Gulf of St. Lawrence region did not fully free Gulf fishermen from the influence of packers. Packers retained significant power by controlling transportation networks and by reorganizing their operations to serve both the canned and live markets. Unlike their counterparts on Nova Scotia's South Shore or in the Bay of Fundy region, Gulf fishermen lacked the option of shipping their catch directly to Boston.

Regional packers, and to a lesser extent fishermen's cooperatives, filled this gap. Pictou-based Maritime Packers, referred to locally as "Broidy's," claimed to be one of the largest shippers of live lobsters in the world from around 1926 until it was sold to National Sea Products in 1965.[101] Maritime Packers began in 1911 when Russian-Jewish merchant Samuel Broidy established his first cannery in River John, Nova Scotia, followed in 1921 by a second at Cape Tormentine, New Brunswick.[102] His son William joined the company in 1923 and took over as general manager in 1927. While other canners retreated from a volatile industry, Maritime Packers expanded into the Magdalen Islands, Prince Edward Island, Cape Breton, and Newfoundland. It opened its own pound in 1927 at Bay View, just outside Pictou, to export live lobsters from the Nova Scotia side of the Northumberland Strait. In 1936 Broidy moved his operations to Pictou where a large facility at nearby Caribou let him buy and process both large market lobsters and small canners.[103]

Although never as large as Maritime Packers, Fred Magee's "Mephisto Brand" canned lobster followed a similar trajectory. Launched in Port Elgin, New Brunswick, in the late 1890s – after the industry's peak – Magee eventually operated three factories in New Brunswick and one in Nova Scotia. He diversified his operations into smoked herring, canned peas, and even can manufacturing, and, like many packers of his era, he also pursued a parallel career in politics. In the 1930s, while many other canneries were closing, Magee gambled on expanding a factory in Pictou. By 1937, he held the perhaps dubious distinction of being the largest packer of canned lobsters in the world – just as the commodity was rapidly losing market value.[104] After the Second World

War, Fred Magee Limited and Maritime Packers collaborated in market segregation and consolidation. Magee's company bought the smaller, canner-sized lobsters from Maritime Packers Limited, while Maritime Packers purchased larger, market-size lobster from Magee. As a canner in a declining market, Magee moved to eliminate competition by acquiring the last regional presence of the once dominant Burnham and Morrill – not for its facilities but to absorb its fishermen and secure them as suppliers. Ultimately, Fred Magee Limited itself was acquired by New Brunswick's Paturel in 1963.[105]

Émile Paturel of Cap-Bimet, Shediac, was established by a native of Saint Pierre and Miquelon who came to New Brunswick via France and the United States and began packing lobsters before the First World War. In a similar cohort to Pictou-based William Broidy of Maritime Packers and Fred Magee, Paturel made Shediac a centre of lobster distribution for Southern New Brunswick, northern Nova Scotia, and Prince Edward Island, with exports to the United States.[106]

As profit margins on canning grew very thin, packers looked to other sources of revenue. In 1934, for example, it was estimated that the cost of producing a case of canned lobster in half-pound tins was $22.58 and the selling price was only $23.30.[107] It therefore makes a great deal of sense to see diversified revenue streams on Maritime Packers' 1930s letterhead that advertised "Gasoline and Motor Oils" along with "Live and Canned Lobsters." Its business was not only buying and processing lobsters but continuing the tradition of selling essential supplies to fishermen. But unlike the days of canning monopolies, lobster buyers in the 1930s increasingly faced competition themselves. A Maritime Packers calendar from the decade – likely a novelty item for promotional purposes – advertised the company as "Packers and Exporters of Lobsters, Alive and Canned," with the slogan "A Fair Deal for the Fisherman and Fair Price for His Fish."[108] Other lobster buyers attracted fishermen who now had a choice in where to sell their catch by offering incentives such as discounted gasoline prices.

The live lobster business was unquestionably overtaking the canned sector. Between 1936 and 1939, the number of licensed canneries tumbled from 256 to 191, a drop of nearly 25 per cent, with the Magdalen Islands and Prince Edward Island being particularly hard hit.[109] Moreover, regional packers increasingly had to contend with the growing presence of American buyers in the live market. No longer content to anchor their smack boats offshore, these buyers – primarily based out of Boston or Gloucester – became influential players, particularly as transportation improvements brought Cape Breton and the west coast of Newfoundland into the live trade. One of the most significant of these firms was James Hook & Company of Boston which was established in 1925.[110]

Among the most notable Boston-based figures in the interwar lobster trade was a woman: Fanny Nickerson Powell, daughter of Moses H. Nickerson. Raised on Cape Sable Island, Fanny grew up immersed in the lobster industry. Her father, a prominent figure in both the canning and later live market sectors, was also a journalist and Liberal politician who played a central role in the founding of the Fishermen's Union of Nova Scotia – an early association representing boat-owning fishermen that focused on prices, operating costs, and political advocacy. Like many Nova Scotian young women of the time, Fanny moved to Boston, where she initially worked as a bookkeeper for the Boston Lobster Company. In 1908 she married her employer, Avery Powell, but later struck out on her own after he joined the Boston Lobster Association. Drawing on her Canadian lobster connections, she founded the Consolidated Lobster Company based in Gloucester, opening a second branch in Chelsea by 1931. Her success was considerable: in her first six month of operation, she reportedly earned $25,000 – a figure so striking that her husband eventually returned to work for her company.[111]

Amid the economic hardship of the Depression Ottawa sought ways to help fishermen on Nova Scotia's Eastern Shore access Boston's lucrative live market. To that end, the federal government subsidized a collection boat to transport crated lobsters to Boston at significantly lower freight costs than those offered by the limited private operators. Private boats charged as much as $6 freight per crate, but the government-subsidized vessel offered the same service for $3.[112] This initiative culminated in a notable shipment in May 1933, when the coastal freighter SS *Dominion Halsyd* arrived in Boston after a three-day voyage from Isaac Harbour, carrying 536 crates of live lobster on board.[113]

With the expansion of the live lobster market came a change in government policy concerning holding pounds. Lobster pounds were a contested political issue in the 1890s and 1900s. They also remained a constant source of frustration to Canadian fishermen, who knew that many of their live sales were being held in pounds in Maine. If these lobsters were carrying eggs, they were believed to be adding to Maine's breeding stock instead of Canada's. In 1928 Edward Leonard Conley was granted a licence to operate a pound on Deer Island near St. Andrews, New Brunswick, and by the end of the Second World War Conley claimed to be the world's largest independent lobster shipper, handling over 15 per cent of the Canadian catch.[114]

Generally, prices fluctuated throughout the year according to supply. Usually, prices were highest in late March or early April when only the Fundy and South Shore fishermen were legally fishing and before the first Gulf areas opened. Live lobsters had to be sold immediately, as most fishermen did not have the capacity to hold them alive until prices peaked. Prices could also change very quickly. In April 1932, large lobsters had been receiving 35¢ a pound (or $42 a crate) in Boston, but when 800 crates arrived in the city from Nova Scotia, the price dropped immediately to 20¢ a pound or $28 a crate.[115] But even the lower price must have been a source of envy and frustration to fishermen in Prince Edward Island who that summer only received 3¢ a pound for the canners they sold.[116] Only five months earlier, in December 1931, the *Halifax Chronicle* reported that fishermen were receiving the lowest prices in twenty-five years. This was the result of a combination of the loss of a consumer market for a luxury good and the surplus of "one million pounds of lobster," caught that fall and held in cold storage. It was a disaster, as low prices in the 1930s coincided with the intensification of fishing efforts. With few employment opportunities, the end of work on the salt bank schooners, and immigration to the United States closed, the lobster fishery acted as an employment of last resort. The *Halifax Herald* reported that the number of lobster fishermen had increased from 10,000 to 17,000 between 1929 and 1937.[117] In the Gulf region, where canning persisted, most fishermen through the interwar period continued to fish with company gear. Fishermen signed contracts in the spring that set prices, and they settled twice during the season. They never completely cleared the debts on credit advanced, so stayed on the company's books year-round as the cycle of credit and debt never ended. Their lack of boat ownership meant that in some fishing communities they were dismissed by others who earned their living on the water as "not real fishermen."[118]

In the independent fishery, boat owners themselves would hire at least one helper to aid with hauling traps. The helper would usually work for a share of the catch and was often in a very vulnerable position. In the French-speaking Pubnicos, times were so tough in the 1930s that a corruption of the English word "beggar" became "begou" as the local word for helper.[119] With the price of gas higher than the value of labour, there were reports of certain fishermen replacing their motors with oars.[120] Glimpses of this squeezed workforce were also evident in the 1937 tragic drowning of a father and son at South West Port Mouton. Here a poor season had meant that the forty-three-year-old father replaced his hired man with his fourteen-year-old son.[121]

Figure 9. Wartime poster, *Canadian Home Journal*, December 1939. The Canadian government engaged in a domestic marketing campaign to dispose of an overstock in canned lobster.

But hard times also encouraged new specialized side industries. The most famous was in West Pubnico in the 1930s, where tiny wooden pegs used to immobilize claws to prevent damage during shipping were produced for $1.25 per thousand.[122] The expanded live market sector also required enormous quantities of ice.[123] Refrigerated trucks used an estimated three pounds of ice for every one pound of lobster shipped.[124] Branching out into the wooden peg, ice, or even bait business were new or expanding schemes to profit off the growing sale of live lobsters.

The Second World War created a sudden emergency for the declining canned sector when Great Britain, the market for 70–75 per cent of the pack, again declared a wartime embargo on the importation of the luxury good.[125] To compensate for the loss of this market, the Canadian Department of Fisheries purchased the overstock, set up the Lobster Control Board under the War Measures Act, and aggressively advertised canned lobster to Canadian women. Through adding canned lobster to their grocery order, housewives could be patriotic and prepare delicious family meals. The advertisement campaign may have had some success as the Canadian domestic market increased two and a half times. While some Canadian households may have added lobster to the menu, historian Brian Payne points out that canned lobster was sold to the Canadian military as camp rations and "a very large percentage of government stock" was sold "to the government itself."[126] Americans also took up the excess supply, and by 1940 exports to south of the border had doubled. By 1942–3, there was sufficient North American demand that Canadian government special assistance to the industry was withdrawn.[127] Although the immediate crisis was addressed, the fate of canned lobster was sealed. The war also played havoc on those exporting live lobsters to the United States. Issues with the exchange rate and the reallocation of transportation links to war service challenged fishermen as much as gasoline rations and shortages of rope.

Post-war

By the end of the 1930s, the value of live lobsters surpassed that of canned lobster, and in the post-war period, lobster processed in its traditional form became a less important market share. The wartime obstacles that had limited the live market trade receded with peace. By 1946 the value of live lobster was roughly three times that of the canned product, and by 1961 live lobster was about 90 per cent of the value of lobster exports.[128] Lobster as a fresh or frozen product remained primarily an export product, and this meant an almost exclusive dependence on the United States. In 1956 over 96 per cent of non-canned lobster exports went there.[129] The development of this market reflected advances in transportation (the first experiments with lobster airfreight followed the war) and refrigeration. It also reflected consumer preferences and a rising standard of living, with some middle-class consumers mimicking Hollywood 1950s glamour with opulent dishes such as lobster thermidor, frequenting restaurants with live lobster tanks, or embracing the growth of Chinese food restaurants in Eastern US cities.[130] Live lobsters also received a local domestic boost in consumption as regional markets, associated with the growth of automobile-based tourism, developed. The Pictou Lobster Festival began in 1934 and was joined by the Shediac Lobster Festival in 1949 and the Summerside Festival in 1955. These events were organized by local businesses and served local communities but attracted large numbers of visitors. Lobster suppers were a traditional form of community fundraising, even if they had created awkward situations for legal enforcement. Church fundraisers in Yarmouth County in the early 1930s accepted donations of illegal undersized lobsters with the full support of local priests that left local fishery officers' hands tied.[131] In the 1950s, on Prince Edward Island, these sporadic community-based events were transformed into regular fundraisers directed specifically at tourists, such as the New Glasgow and District Junior Farmer's Organization's supper established in 1958 and the St. Ann's Catholic Church, Hope River supper started in 1964. The St. Ann's supper began offering lobster one night a week and ended up operating six evenings a week from June through September until 2015.[132]

In a period when the offshore fin-fish fishery was industrializing with enthusiastic support of the federal government – particularly following Newfoundland's entry into Confederation in 1949 – the inshore lobster fishery was cast as old fashioned, even irrelevant to the future of Canadian fisheries. It attracted minimal corporate investment, saw few technological innovations, and remained largely unchanged in practice. Lobster gear evolved little, and the exclusively inshore fishery remained relatively low capital, labour intensive, and decentralized. It was quaint and picturesque, but not completely untouched by new industrial processes. By 1946 Maritime Packers had shifted away from producing traditional canned lobster, processing instead only fresh and frozen meat. From 1948 to the mid 1960s, the company packaged fresh lobster meat in twelve-ounce cans with cellophane tops that were shipped weekly to the United States. While Maritime Packers continued to sell canned lobster under its "Maripac" brand, the product was no longer their own – it was purchased from sub-suppliers and merely relabelled.[133]

Eventually, these changes led to consolidation of the processing and distribution sectors. Around 1950 the last of the old American canning companies shed its interests in packing, and in August 1961 the Consolidated Lobster Company of Gloucester closed, causing a crisis in the Newfoundland live lobster market.[134] The early 1960s witnessed a period of industrial consolidation that in 1965 resulted in the Lunenburg-based National Sea Products acquiring Maritime Packers of Pictou, E. Paturel of Shediac, and St. Andrews Conley's Lobsters.[135]

Figure 10. Magee's canning factory, Pictou, Nova Scotia, c. 1946. A gender-segregated labour force persisted as women continued to extract lobster meat by hand. The process remained difficult to mechanize.

www.novastory.ca/cdm/singleitem/collection/phps/id/1842/rec/2
With permission of the Pictou Historical Photograph Society

Lobster no longer preoccupied the federal government or business interests, but this "traditional" decentralized fishery continued to offer, at least in relative terms, high and assured returns for individuals who fished. After the war, in March 1946, lobster fishermen in Southwest Nova Scotia, including newly returned veterans who had access to boats, were receiving 85¢ a pound for live lobsters, with prices dropping to between 40¢ and 50¢ a pound over the summer in the Gulf region.[136] These strong prices encouraged some veterans to enter the fishery at least temporarily. Certainly, there was more money in lobster than in other fish. In 1951 lobster landings comprised 81 per cent of Prince Edward Island's total landed value of fish compared to only 12 per cent for Newfoundland, with New Brunswick and Nova Scotia falling in the middle with 51 and 43 per cent respectively.[137] This put money in the hands of fishermen. In 1953 Maritime lobstermen earned $14.5 million for a catch of 42 million pounds compared to the cod fishermen who earned only $5 million for their catch of 183 million pounds.[138] As a high value catch, through the late 1950s, lobster provided triple the income to fishermen in the Maritimes compared to cod. As the lobster prices paid to fishermen increased between 25 and 30 per cent between 1955 and 1965, there was a dramatic increase again in the number of fishermen engaged in the fishery. In 1963 23,500 lobster licences were issued. A representative of the Maritime Fishermen's Union claimed that this number was a 40–45 per cent increase from 1955 and that an increasing percentage of fishermen were pursuing lobster on a part-time basis. Since regulations required all lobstermen to engage on a restricted basis, this depiction was meant to draw attention to increased participation by those who were not so-called bona fide fishermen but were engaging in the older tradition of occupational pluralism.[139]

Although lobster generated important revenue, the standard of living of lobster fishermen remained low and was at odds with postwar prosperity. The increasing costs of boats meant that it became more expensive to enter the fishery, especially as processors gradually abandoned boat ownership and "fishing the company fleet" became a thing of the past. Historian Peter Rider estimated that in the 1900s, with $100 a fisherman could enter the lobster fishery at a time when $350 was an annual living wage. In comparison, by the 1950s, $2,700 was needed for a boat, engine, and a season's gasoline, an amount that exceeded a minimum annual income of $2,500.[140] Lobster fishermen's poverty was a preoccupation by many involved in public policy. Research conducted by the St. Francis Xavier University Extension Department among fishing households in Guysborough County in 1965 revealed that almost three quarters of households spent at least half of their annual income on food, an amount much higher than the national estimate of 30 per cent.[141] Fishermen in the Gulf and Guysborough County were not necessarily independent commodity producers, and the director of the St. Francis Xavier Extension Department described the 1950 situation of fishermen on the Antigonish shore as being akin to "sharecroppers for big lobster packers," as in exchange for "outside gear and boats" they consigned their lobsters to "certain" packers in New Brunswick and PEI.[142]

The number of traps individual fishermen tended steadily increased as yields per traps decreased. In 1880 the average fisherman fished 75–90 traps, but by the end of the First World War this number had risen to between 250 and 300 traps. In 1912 packer Munro McKenzie of

River John, Nova Scotia, wrote to the Minister of Marine and Fisheries and claimed that some fishermen were running up to 500 and even 600 traps.[143] Newfoundland lobster fishermen, part of the Canadian market after 1949, fished less intensively. In 1963 lobster fishermen on the south coast of Newfoundland were still only fishing in six to eight fathoms (around 11–14.5 metres) of water as they hauled their traps by hand rather than use the mechanical haulers long employed elsewhere.[144]

Once fishermen became eligible for Unemployment Insurance in 1957, this income support program influenced both employment and fishing strategies.[145] As will be discussed in chapter 7, fishing intensity now took into consideration eligibility requirements to qualify for seasonal benefits. With a regulated short fishing season and relatively low individual capital investments, it continued to make financial sense for fishermen to invest their sweat equity into making and repairing their own traps, rather than use sparse cash to buy the new manufactured traps that were becoming available. Regulations concerning gear worked together with personal decisions that weighed the trade-off of much time and little money to encourage self-sufficiency.

Conclusion

"Canadian" lobster's transformation into a commodity, from an abundant, low-value, local household or community food source, a fertilizer, or bait for other fisheries to a canned modern industrial product that could be safely (and tastily) preserved or kept alive and shipped long distances as a luxury food, shaped the fishery. Lobster took on value as it entered the capitalist marketplace. The related industry was circumscribed by the ecology of lobsters themselves – their habits, the decentralized distribution of their habitat, their reproductive cycles, and their seasonal suitability for surviving long-distance shipping. These environmental traits interacted with the region's geography and the pre-existence of occupational pluralism where a new part-time, widespread, and decentralized fishery worked within existing seasonal economic strategies. The history of lobster as a commodity shaped its worth, its labour force, and the corporate and entrepreneurial opportunities it created as a transnational product. The development of the canned sector and the parallel market for live lobsters shipped to the United States had distinct geographic concentrations and generated different strategies for American, British, and Canadian companies to both make profits or lose money. It also shaped the state as fisheries bureaucrats, politicians, and enforcement agencies developed strategies for protecting a natural resource with regulations and enforcement. Almost at the birth of the canned industry in Canada came state regulations. Law coincided with the rise of lobster's market value. Cast as a valuable national resource for the young nation, the federal government mediated conflicting interests between communities, packers, fishermen, and itself. Lobster politics reflected shifting state strategies connected to a changing commodity and the ever-shifting and diverse interests of the many people involved.

Chapter Three

Regulating the Lobster Fishery

It is almost impossible to overestimate the significance of the lobster fishery to Canada's Atlantic coast. Its importance is not just economic, it is also social and political. The 1898 Royal Commission on the Lobster Industry described this shellfish as the staple industry in many coastal communities and that everywhere was "an important lobster ground." In particular districts, such as the south and east coasts of Cape Breton, the Eastern and South Shores of Nova Scotia, and the Northumberland Strait, as many as 75 per cent of fishermen depended solely on their lobster catch.[1] Ten years later, Dominion Superintendent of Fisheries expressed the challenge of establishing the number of fishermen engaged in the trade as "everybody is a lobster fisherman. The farmer fishes, and apparently everybody in the neighbourhood."[2] In turn, the broad geographic and economic footprint of the lobster industry amplified its political influence. By the turn of the twentieth century, the fishery was economically important in almost all of the more than thirty federal ridings bordering the Atlantic Ocean. The creation of the federal Department of Marine and Fisheries following Confederation in 1867 coincided with the establishment and rapid growth of the commercial lobster fishery. As a result, the introduction and enforcement of lobster regulations became one of the earliest and most tangible manifestations of Dominion authority in coastal communities – potentially marking one of the first ways people in coastal communities encountered the new federal state.

While the department responsible for fisheries was predominantly preoccupied in the 1860s and 1870s with international issues such as protecting Canadian sovereignty against American encroachment – it also attended to local matters.[3] It positioned itself as a neutral mediator in disputes between fishermen and canners, while promoting the conservation of natural resources as a foundation for national economic development. Yet even a young and relatively weak state was never without its own interests. The new Dominion government built upon pre-existing colonial fishery conservation legislation for established valuable fish such as Atlantic salmon. Lobster, however, had not been fished commercially, and it was one of the first fisheries that Ottawa addressed from scratch. There was always a small subsistence lobster fishery for both Indigenous peoples and settlers. For European settlers, it was generally outside the market economy and so had limited economic relevance. As it had not been fished intensively by any group, lobster's sudden dramatic change in fortune had no precedent of informal local customs governing access, or what historian Karl Jacoby has referred to as "non capitalist forms for ecological management."[4] Lobster fishing was without an older tradition of specific settler folk regulation rooted in a moral economy, and local informal community practices governing intensive fishing emerged in tandem with the market and state regulations. Historian Brian Payne's observation that Maine sardine fishermen, another fishery transformed by preservation technology and industrialization, did not see "conservation and commercial production as mutually exclusive" can also be applied to many who fished for lobster.[5] The community-based practices that developed under capitalism were not oriented solely towards the market and reflected local cultural practices and values. Perceptions of the commons had very local contexts.

Decisions about an intensely local fishery were made 1,500 kilometres away from the people or the lobster affected. Time and time again, the attempt to regulate the lobster fishery revealed the federal state's efforts to make up for organizational failings at the local level with overly centralized decision-making.[6] In order to understand the decisions made – biological, economic, political, and social forces must be brought into play. Lobster regulations were the result of the politics that intersected at the crossroads of technology, local fishing, the environment, and the market. How people fished shaped regulations, and in turn regulations shaped how people fished. Codification of local practices was a less coercive strategy for establishing and maintaining even tenuous order and so formal and informal laws coexisted. This chapter explores how after 1873 formal state regulations worked together and in conflict with informal practices to police the lobster fishery.

The regulation of the lobster fishery was structured both by formal legislative laws and regulations and informal customs and traditions, those "unspoken laws" that governed community practices. The potential coercive power of the state was mirrored by local direct action, which took the form of cutting trap lines, destroying traps, or worse.

Formal legislation reflected the economic and social importance of protecting a valuable and vulnerable ocean resource. But at least until after the Second World War, the impact of regulations was severely limited by the Canadian state's capacity to govern and its capacity to know.[7] Strategies around formal legislation must also always be considered within the context of the New England experience, both as creating the dominant economic circumstances and as the adjacent ecological domain.

Official lobster regulations emerged concurrently with the new Dominion of Canada. A great deal of fishing took place outside of any judicial jurisdiction, as the Canadian state only claimed territoriality of fishing grounds three nautical miles from all headlands until 1964. That year an exclusive fisheries zone was expanded to twelve nautical miles, and in 1977, a unilateral declaration established a 200-nautical-mile Exclusive Economic Zone.[8] But even international boundaries were a constant source of formal and informal conflict. American-operated smacks, which collected live lobsters from fishing communities, regularly fished and anchored outside the international three-mile limit in Southwest Nova Scotia where they were supplied with lobsters caught in international waters by local fishermen in local close seasons. In 1911 there was an international incident as the Canadian government unsuccessfully attempted to stop two Boston-based smacks, the *J. R. Atwood* and *Pride of the Port*, from anchoring just beyond the three-mile limit off Shelburne County and buying lobsters out of season. The federal Minister of Justice regularly reminded the frustrated Department of Marine and Fisheries that it had no authority to interfere. Around the same time in Prince Edward Island, before the open season began, fishermen visibly set traps outside the three-mile limit with no legal consequences.[9]

While the Canadian state claimed jurisdiction within tidal water, English common law upheld a public right to fish derived from one's status as a British subject and the doctrine of common interest. Common interest holds that air, rivers, sea, and seashore are held in trust by the Crown as a public resource. Public access to these resources can be limited by Parliamentary legislative restrictions to preserve and protect the resource. Originally both inland and sea-coast fisheries were exclusively a federal power supervised through the federal fisheries department. In 1898, however, the Judicial Committee of the Privy Council in London, England, ruled that the "property rights" of saltwater fisheries were those as defined at the time of Confederation, unless they were transferred by expressed enactment.[10] Parliament, or after 1898 in the case of Quebec, the provincial legislature, could limit the conditions through "legislative rights" related to conservation, but it could not exclude any British subjects from the right to fish in tidal waters; access was a common property.[11]

This confused and often overlapping jurisdiction between the federal and provincial governments became somewhat clearer in a 1913 ruling, which established that public access to the fishery was distinct from provincial proprietary rights over non-tidal waters. From 1921 to 1984, the federal government granted Quebec authority to administer its own sea fisheries through its Ministère des Pêcheries – with the exception of the Magdalen Islands, which remained under federal control until 1943. Other provinces generally accepted federal jurisdiction over fisheries beyond the tide line or the wharf.[12] Above the tide line was another matter. In 1929 the Judicial Committee of the Privy Council upheld a decision by the Supreme Court that the federal government had no right to control canning licences, as once landed, fish entered a provincial domain. Ocean fish, therefore, changed from federal to provincial jurisdictions where the provinces took over responsibility for licensing processing plants. To muddy things further, the same fish or fish product re-entered federal jurisdiction by crossing a provincial border or as an international export.[13]

In 1873, after three years of study and consideration by W.H. Venning, federal Fisheries Inspector for Nova Scotia and New Brunswick and W.F. Whitcher, Dominion Commissioner of Fisheries, Ottawa began an active program of government oversight and intervention that restricted the legal ability to fish lobsters bearing eggs, created geographic close seasons and size limits, and eventually moved into regulating gear.[14] The regulations came from self-taught practical fisheries bureaucrats who combined their field experience with what was happening in Maine and Massachusetts. Claims to authority were broad and elastic. Unlike Newfoundland, where the British maintained that lobsters were not fish, and therefore that "catching and canning lobsters were not the same as catching and drying cod," this was never a question before Canadian courts.[15] Generally, the history of this regulatory regime is a history of the state's incapacity to enforce various regulations and the challenge it held in gaining cooperation from fishermen, packers, and local politicians. The very first legislation in 1873 was met with organized opposition from packers, and this was a constant theme in what was more than once described as a three-corner fight between packers, fishermen, and the state. Although there were formal regulations, they were frequently not understood, known, or accepted at the local level. On occasion fishermen wrote directly to the Department of Marine and Fisheries for clarification. In 1911, for example, a fisherman from Bonaventure, Quebec, asked the Department if a local packer could prevent him from setting traps.[16]

Across the Gulf of Maine in the United States, where sea fisheries were under state jurisdiction, legislators began introducing modern conservation laws as early as 1868. That year, canners were required to buy lobsters based on weight rather than count. In 1872, Maine regulations prohibited the buying or selling of berried lobsters (females carrying eggs), and by 1874 canneries were required to close from 1 August to 15 October. These measures proved insufficient to stop the population decline. In 1879, the close season was significantly expanded to run from 1 August to 1 April. Finally, the introduction of a year-round size limit in 1895 rendered the few remaining Maine canneries unviable, bringing that phase of the industry to a close.[17] In the early 1930s Maine adopted the "double gauge" law that imposed size restriction for lobsters that were classified as both too small and too large, in the hope of protecting both smaller lobsters and the larger breeding stock.[18] The practice of tail-notching egg-bearing lobsters, marking them with a "V notch" to indicate they should not be caught, was adopted informally in the late 1930s.[19] Maine's laws influenced the Canadian fishery both as models to accept or reject and shaped the Canadian industry's existence and viability. At the level of government, science, and business there was a mixture of cooperation, integration, and competition, and the "Canadian" lobster fishery cannot be understood without taking the US market, ownership, and state politics into account. Examples of transnational cooperation in lobster regulation can be seen in meetings, such as the 1903 conference in Boston attended by the Deputy Commissioner of Fisheries, Robert Norris Venning and A.C. Bertram, Inspector of Fisheries for Cape Breton.[20]

Canada's first lobster conservation legislation, enacted in 1873, protected recently moulted (soft-shelled) lobsters, lobsters weighing less than one and a half pounds, and berried lobsters. These regulations were revised the following year: reference to soft-shelled lobster was replaced by a close season in July and August, the prohibition on berried lobsters remained, and size limits were now based on length rather than weight – specifically, a minimum of nine inches head to tail. Recognizing the vast ecological differences between the Atlantic coast and Gulf of St. Lawrence, the federal government divided the fishery into two separate districts in 1877, each with its own close season. The first district covered the Gulf of St. Lawrence adjacent to Quebec and New Brunswick (closed mid-August to mid-September) and the second the

Figure 11. Map of lobster fishing districts following the 1898 Royal Commission.

coast around Nova Scotia, Prince Edward Island, and New Brunswick's Bay of Fundy, which was closed to legal fishing for August. The dramatic decline in the catch expanded the close season from four weeks to eight months in 1879 for both districts. This spatial division eventually shifted from being organized primarily around provincial land boundaries to ones based on marine geography. By 1887 the fishery was divided into two main districts: one stretching from the Maine/New Brunswick border to Canso, and the other encompassing Cape Breton and the Gulf of St. Lawrence. Six lobster fishing districts were introduced in 1899, establishing the general areas that exist today. These districts had close seasons between five and a half months to nine and

Figure 12. Map of 1914 lobster fishing districts, which were a slight revision of the 1910 divisions.

a half months and were based on environmental factors such as water temperature, ice conditions, and the lobsters' reproductive patterns but were connected to complimentary and competing economic activities around farming and other fisheries. The number of different fishing areas with distinct close seasons and eventually different size limits continued to grow so that there were ten in 1910, fourteen in 1940, and sixteen in 1949 after Newfoundland joined Confederation.

Seasons were pegged not only to the lobster's habits but also to fishermen. Local open seasons were established so as not to interfere with other fisheries such as mackerel or cod, or conflict with other means of subsistence such as planting. The lack of a fall fishery was about the quality of product for both the live and canned sectors and the difficulty of getting bait at that time of year. Soft-shelled lobsters caught after moulting did not travel as well as spring hard-shell lobster, and they were not always full of meat. The variable close seasons were also a form of early supply-side management. The rotating seasons controlled the volume of lobsters on the market at any time in the hope of keeping prices as high and stable as possible.[21] The relatively short open season created a part-time fishery from almost the moment lobster had value as a commodity.

Fishing regulations were generally limited to rules around season, size, gear, and who had the right to fish. The most extreme action was the repeatedly floated strategy of a moratorium to save the lobster stock.[22] This was a tactic adopted in independent Newfoundland for the 1925, 1926, and 1927 seasons but seriously divided Canadian fishermen, canners, and politicians. It was proposed before the 1898 Royal Commission on the Lobster Industry, and the Department of Marine and Fisheries subsequently surveyed all regional government Members of Parliament on their opinion regarding a complete closure to get a sense if it might have political support. Policy and politics were always entwined, and it did not. At the end of the First World War, the Deputy Minister of Naval Services, temporarily responsible for fisheries, proposed that he would abolish legal canning during the 1919 season and thereafter restrict it to biannual years for stock recovery. This strategy would not interfere with the live lobster trade but was intended to preserve Canadian international markets for canned goods and the value of investments in factories and fishing gear.[23] It was vehemently opposed by both large and small canning concerns in an organized and successful campaign by packers who argued that it was unfair that the live market would be permitted to continue considering the large amount of capital canners had invested in the region. Big companies disingenuously expressed concern for the small independent packers who would be forced to close permanently and more truthfully worried that they would lose their already transient and temporary workforce.[24] The federal bureaucrats responded in 1923 with a new strategy to keep the sector open but reduce its size by mandating expensive sanitary equipment and uniform standards. In order to get a licence, every packer was graded at the start of the season on a set of criteria. Canners had three years to correct deficiencies and score 100 points, easily achieved by acquiring an expensive piece of equipment, the steam retort, for sterilization.[25] This strategy particularly affected Prince Edward Island where in 1922, 70 per cent of the total value of the

Island's fisheries had come from canned lobster. The Canadian Fisheries Association, which represented the business side of fisheries, had long advocated that industrial consolidation was the best means to raise uniform quality, while acknowledging that such standards would require many smaller canneries to shut down. In the 1930s, when economic conditions already were closing factories, it was politically acceptable to demand that as of the 1936 season steam retorts would be mandatory in all export packing facilities.[26]

The Dominion government attempted all kinds of strategies to enforce its regulations – a task especially challenging in a context where it possessed virtually no coercive capacity nor, at least in the early years, much of a presence in most parts of the region. In the late 1890s, the federal government looked both south and east for inspiration. The division of federal and state powers in the United States meant that the American federal government had no fishery enforcement rights at the local level and therefore had limited options in conservation. As a result, in the case of lobster, the US federal government turned to propagation to increase the stock. This was a politically appealing option to Canadian politicians, as it was easier to promote the establishment of hatcheries, with their related patronage positions, than unpopular regulations. The Dominion of Newfoundland had entered the propagation field in 1891 and shortly after through the First World War, as will be discussed in chapter 4, it became central to Canadian federal lobster policy.[27]

In the 1890s the Dominion government under new leadership introduced controls through licensing canners. The new Commissioner of Fisheries Edward E. Prince arrived in 1892 and identified the licensing of canners for both lobster and salmon as essential for the regulation of both fisheries. Adherence to the law might be leveraged through the threat of a licence's cancellation. Nearly thirty years later Prince recounted the opposition he encountered from within the department, as "older department officers" held that "canning was not a fishing operation, and lobstering was not "'a fishery in law.'"[28] This legal definition of lobster referred to its regulation by Orders in Council rather than acts of Parliament. Prince was able to achieve his goal and gradually expanded and codified all lobster regulations. After 1894 packers were required to hold a valid annual licence that required them to label their cans with their names and the year of production.[29] When licences were introduced, all existing canning operations received them. But five years later, in 1899, when the large number of canning operations were thought to be harming the fishery, the Department refused new licences unless they were proposed by fishermen's cooperatives. Any British subject could fish, but not everyone could can.[30]

Licensing canners served different purposes over time. Initially, the primary goal was to curb illegal packing through the use of official stamps, which allowed authorities to monitor production and transportation. By 1900, however, licensing had shifted towards limiting entry into an overcrowded industry. In 1922, regulatory authority over canned lobster and cannery operations was brought under the federal *Meat and Canned Foods Act* and packing licences were used not only to ensure product quality but also to pressure substandard factories out of business.

Licensing fishermen followed during the First World War. After 1 October 1918, any British subject fishing lobster was supposed to obtain an annual licence from the Department of Naval Services for the nominal fee of 25¢, a sum that remained in place until the 1960s. In Quebec, licences were granted by the province at the cost of 1¢ per trap fished, with a minimum of 25¢ per licence. As with the canners, the intent was to encourage compliance beyond fines or jail, as law-breaking licensed fishermen had the potential of having their licences cancelled. Licences were not transferable and had to be produced if requested. They also held the added benefit of monitoring the number of fishermen legally engaged in the fishery. In the 1930s individual licences also became a means of restricting fishermen, and eventually boats, to catching lobster in a single lobster fishing district. In the Gulf of St. Lawrence, where canners were more likely to own (or at least supply) boats and gear, this regulation was aimed at large packers who moved men, boats, and gear around as the open seasons changed. Licences were attached to individual fishermen, and the Department of Fisheries acknowledged in 1961 that it was impossible to enforce the movement of traps, rope, and buoys. Moving boats remained illegal, but the law was flaunted, as the boats in use were the same size and shape and identification in court was difficult or impossible – especially if another colour of paint was applied between seasons. Increased legislation directed at fishermen followed the Second World War. In 1953 the new *Coastal Fisheries Protection Act* required all boats to "heave to" when signalled by a law enforcement boat or a protection officer. It also became an offence – punishable by heavy fines – to throw anything overboard once the signal had been made, making it more difficult for fishermen to dispose of evidence of illegal fishing.[31] As a local fishery conducted almost entirely within sight of land, lobsters – like oysters – were seen by the regulating department as well-suited for management.[32] But this reflected a kind of bureaucratic hubris: in reality, inshore waters could rarely be effectively policed or physically controlled.

Attempts to regulate the lobster fishery were also made by restricting the kinds of gear that could be legally employed. The first effort by the Department of Marine and Fisheries was unsuccessful when in the 1890s it attempted to regulate fishing gear through setting a minimum distance between laths on traps. The proposed restriction of lath spacing to at least an inch and a half apart generated a backlash. A Charlottetown merchant complained that the Fisheries Branch "must think that our lobsters are as big as rabbits," as this gap was so large that no lobsters in Prince Edward Island would be possibly caught.[33] Packers in the Gulf of St. Lawrence who typically owned traps their employees fished, petitioned that since every trap had to be replaced every three to four years, this law should be phased in so that existing traps did not need to be destroyed or remade. Indeed, for large packers such as Cape Breton's H.E. Baker, who claimed he owned between 9,000 and 10,000 traps, this was a considerable expense. Implementation of the regulation was supposed to be delayed until 1897, but the new Liberal government did not enact this regulation until 1910.[34] This legislative intervention into the type of gear employed suspended the basic method for catching lobsters in the early twentieth century, notwithstanding dramatic technological changes brought by motorboats, mechanical haulers, and bottom-sensing radar. Lobsters became exceptional in the modern fishery, as they had to enter the traps in search of bait, rather than be passively snatched up by nets.

Fishermen and packers were informed of regulations through advertisements in local newspapers. The *Cape Sable Advertiser* on 21 July 1887 publicized the date when factories must close and all shipping must stop, but not everyone had access to this information. Regulations during the 1920s took the form of large posters, presumably placed on public wharves and at post offices. A poster for the 1923 season had in bolded, large, capitalized, and underlined typeface at the bottom: **THOSE FOUND GUILTY OF DEFACING THIS POSTER WILL BE PENALIZED**.[35] The fact that this needed to be stated suggests that not everyone appreciated the regulations or their publication.

Federal fisheries officials regularly experimented with regulatory strategies through "special permissions" almost always associated with pre-existing political connections. This allowed the department to pilot new strategies and simultaneously reward its political friends. One of the challenges those in the live market sector faced was the perceived unfair advantage that Americans had through their use of holding pounds, which allowed them to store live lobsters until prices were highest. This was particularly infuriating, as American smack

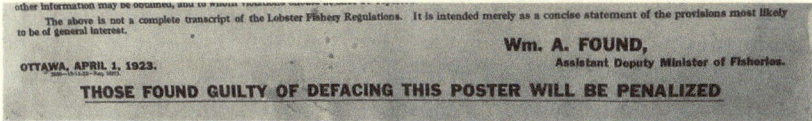

other information may be obtained, and to whom violations thereof...
The above is not a complete transcript of the Lobster Fishery Regulations. It is intended merely as a concise statement of the provisions most likely to be of general interest.

Wm. A. FOUND,
Assistant Deputy Minister of Fisheries.

OTTAWA, APRIL 1, 1923.

THOSE FOUND GUILTY OF DEFACING THIS POSTER WILL BE PENALIZED

Figure 13. Close-up lobster season notice poster 1923, Wallace and Area Museum, Wallace, Nova Scotia.

boats bought Canadian lobsters, carried them across the Gulf of Maine, held them in their own storage facilities, and then sold them at prices substantially higher than the original cost. In Canada, shipping and holding live lobsters outside the local legal season was prohibited, as it was impossible to monitor when and where the shellfish were originally caught. Experiments with special pounds were attempted in the early 1890s with the Robert E. Harris Fish Company at Long Beach, Digby County, Nova Scotia. Here a fifteen-acre pound stored 60,000 live lobsters and shipped them to the United States between 15 July and 15 August after the local season had closed but when market prices peaked.[36] Opponents of this pilot program observed that many of these exported lobsters were berried (and so Canada was supplying Americans again with valuable lobster eggs) and that the facility itself was not suitable. When the Dominion Commissioner of Fisheries visited the pound in the middle of winter, he found thousands of dead lobsters; the water was not deep enough for the lobsters to survive the cold.[37] The most important obstacle, however, was that as the close season became the most important regulatory tool for conservation, this enforcement was incompatible with available live lobsters.[38] Harris lost the ability to store lobsters with the election of the Liberals in 1896 but eventually recouped his financial outlay with the return of the Conservatives. In 1912 Harris's derelict lobster pound was purchased by the new Conservative Minister of Marine and Fisheries for $10,000.[39]

The federal government experimented with regulatory practices in specific geographic locations. In the 1890s it created a modern fiefdom in the Canadian Labrador, and after 1900 on Anticosti Island something similar emerged with a monopoly grounded in private property rights rather than political connections. There is nothing explicit in the Fisheries Branch records, but it appears as if the Department may have informally followed Newfoundland processing practices in dealing with the Lower North Shore. After the initial boom in the Newfoundland lobster industry, which involved the participation of major New England companies, the greatly reduced and widely dispersed fishery adopted a production model akin to home canning. Just as household labour was used to "make fish" (i.e., preserve salt cod), lobster canning took place at the household level, with local merchants supplying cans and connecting a supply chain to the larger exporting companies. After 1894, when canning licences became mandatory in Canadian jurisdictions, an unusually high number of licences were issued on Quebec's North Shore, despite the relatively low volume of processed lobster recorded. In 1912, for example, each of the forty-seven licensed canneries there produced an average of 40.15 cases per "factory." In stark contrast, Yarmouth County, Nova Scotia – at the other end of the spectrum – had twenty licensed packers who each averaged 721.6 cases per permit.[40]

The unusual context of Quebec's North Shore led the federal government to its most audacious experiment in private licensing. In 1895 Arthur "Gat" or "Gatling Gun" Howard approached the federal government with an idea to solve the local economic crisis that had followed the failure of the cod and salmon fisheries. Howard, who was already operating two "large" canning factories in the area, petitioned the government for an exclusive lease that would give him the sole right to all lobsters and lobster canning in the over 100 kilometres between Ha Ha Bay (near La Tabatière, Quebec) and Blanc Sablon for a term of ten or twenty years, renewable with an amended close season. In exchange, he agreed to employ between eighty and a hundred people and use the offal for fertilizer.[41] Department bureaucrats raised all kinds of objections (it went against the most basic tenants of public access to the fishery), but the Minister approved a five-year lease.[42] Upon this becoming public, a group of Nova Scotians requested a similar monopoly in another area, but they were denied, as this form of resource management was labelled "an experiment."[43]

Howard (1846–1901) was an American who came to Canada "as a friend of" the Colt Firearms Company to operate the Gatling gun, the world's first successful machine gun, loaned to the Canadian militia for use against the Métis in the 1885 Northwest Rebellion.[44] After the Rebellion, Howard stayed in Canada and used his new political connections to gain tariff concessions for his Dominion Cartridge Company. This was the first major commercial munitions factory in Canada, closely tied to Conservative politicians such as its first president (and future prime minister) Sir John C. Abbott and located in Abbott's constituency in Brownsburg, Quebec. Arthur Howard became wealthy as a major shareholder and works manager until 1892 when, while remaining in Brownsburg, he established a mercury fulminate (the chemical used to spark explosives) manufacturing facility in Quebec's Eastern

Townships.[45] By 1894 he had moved on again to the North Shore of Quebec. In addition to his rather remarkable five-year lobster fishery lease on 100 kilometres of coastline, he also acted as local customs officer and magistrate for the district. This legal authority – and his willingness to use force – gave him immense power as he prevented other Canadians from fishing the area and launched an international incident by the use of his "violent conduct and bad language" when he seized the vessels and property of Newfoundlanders who seasonally fished the coast.[46] In a place where the Dominion government had no capacity to govern, Howard enforced his own fiefdom with "a large schooner well-armed with guns, revolvers, etc." The St. John's *Evening Telegram* condemned the judgment of the Dominion government as to how it "could give such a man such power is more than we all can understand."[47] Indeed.

Newfoundlanders, as British subjects, had the right to fish the area as long as they complied with Canadian conservation legislation. R.N. Venning, the Assistant Commissioner of Fisheries, therefore had to explain to a new Liberal Minister that while Howard had "no power to prevent these vessels from fishing in a general sense," he had exclusive lobster rights, so "if any such interference has taken place, it is in the protection of his own interests."[48] The Canadian government did not issue a monopoly licence for lobster again until the 1980s.

The new Liberal government appears to have ended this "experiment," which may have suited Howard just fine, as he was able to redirect his next imperialist adventure at South Africa. The Department of Militia and Defence turned down his initial offer to provide a battery of four machine guns at his own expense, but in his early fifties he accepted the position of machine gun officer in the 1st Canadian Mounted Rifles. When his unit returned to Canada in December 1900, Howard remained to organize and command the Canadian Scouts until his death in February 1901.

Howard was not alone in acquiring exclusive access to lobster and lobster canning. There were parallels between his use of political connections and the property claims that provided Henri Menier, the French chocolatier who purchased Anticosti Island in 1895, with the same privileges. After evicting settlers who had been pursuing their own lobster fishery, Menier established a "model" cannery at Fox Bay in 1901, under the supervision of an experienced Nova Scotian manager. The cannery was run by electric power and employed the most up-to-date international canning practices, pressing tin without solder. Quebec francophone women and men were recruited to work the cannery, while most of the fishermen were English speakers from Nova Scotia. With a monopoly based on property rights that controlled all economic transactions on the island, Menier did well for the first couple of seasons; however, the area was soon fished out and it was not possible to justify the ongoing investment.[49] The experiments associated with Howard and Menier were exceptional and represent extremes in formal regulations that denied local access to the commons through state decree or private property rights through a formal distant law. But the same demands for exclusive access were also made within the framework of informal, everyday, local practices.

Community-based notions of territorial ownership were a persistent theme. On New Year's Day 1912 an anonymous Peggy's Cove fisherman directed a letter to the non-existent Canadian Department of Naval Affairs. The author wrote to complain about lobster fishermen in the next community and concluded, "what we want is a limit between us. Please send the cutter down and have it settled, if not there will be a continual feud between both parties."[50] Explicit in this anonymous request was the belief that the lobster fishermen of Peggy's Cove had distinct fishing grounds where outsiders could be legitimately excluded. Just like the monopolies held by Howard and Menier, it was imagined that the authority and coercive power of the state might be employed to enforce this customary line. There was, of course, no legal basis on which to exclude British subjects from access to a common ocean resource. What we have in this case is a clear example of the state being called upon to uphold a local custom, not a formal law.

Lobster fishermen and the packers who employed them held the idea of a general marine commons based on their "right" to fish as British subjects. But this general principle was tempered by specific contexts as entitlements that were bounded by time and place, and general rights were mediated by local custom that could allocate specific territories to individual fishermen, lobster packers, or communities. This tension was especially acute in certain locales (such as areas abutting different fishing districts with different seasons), places where offshore expansion was not possible, like the bounded Northumberland Strait, and in vicinities and at times when fishermen were earning less money through resource exhaustion or the intensification of fishing efforts. Territorialism was acutely expressed during moments of scarcity or price volatility, as local strategies for protecting exclusive community access were akin to limited monopoly property rights. Local customary areas could be dictated by traditional access, the extension of shore property lines, or shoals claimed annually based on first traps set at the start of an open season. Historian Karl Jacoby's work has emphasized the conflict between the intensely local informal rules and customs of American rural people and their encounter with a distant formal written conservation law, created and enforced by a bureaucratic state.[51] We see the same thing here.

Allegations of illegal fishing were usually directed at neighbouring communities. It happened somewhere else, by others. Typical was a 1903 local news item under "Port L'Hebert News" that accused fishermen in nearby Port Mouton, Nova Scotia, of fishing out of season.[52] Another category of illegal activity was associated with specific places over a long period of time. Certain communities, which did not fall into the normal conflict associated with bordering lobster fishing districts, had persistent reputations for flaunting the law, notably the Egmont Bay area in Prince Edward Island, Marie Joseph on Nova Scotia's Eastern Shore, and Cape Tormentine in New Brunswick.[53]

We know a great deal about territoriality in Maine through the scholarship of James M. Acheson, an anthropologist who has been working in the area since the 1970s.[54] Acheson has developed a portrait of the complicated subcultures and "social networks and obligations" of a highly territorial fishery where lobstermen make location-based claims to communal property that are inherited and protected against trespassers. "Harbour gangs," fishing out of particular harbours, policed specific territorial grounds both for trespassers and adherence to conservation measures. This is an example of lived law: there is no formal legal recognition of communal property and the exclusion of outsiders as trespassers is enforced not by the state but by threats and surreptitious violence.[55]

But the historical condition and circumstances of lobster fishermen in Maine are surprisingly different from that of the Maritimes. In Maine, after the early collapse of canning, all lobster fishermen were independent commodity producers involved in supplying the live lobster trade. Moreover, in the twentieth century Maine lobstermen did not fish from public "government" wharves and only exceptionally were subjected to restrictive close seasons.[56] In Canada, there were fundamental economic and legislative differences. The enduring importance of packers in some regions until even after the Second World War meant that certain firms held tremendous influence as monopoly buyers, suppliers, and sometimes employers. Federal control established set geographic districts, which determined open and close legal seasons. Conflict, transgression, and usage claims were characteristic of spatial and temporal boundaries, and tension was heightened during moments of resource scarcity, price volatility, or intensification of fishing efforts.

If lobster fishing in Canada was premised on the idea of the commons, it was never a completely open commons. Claims to ownership could be as varied as the water directly in front of canneries, literal extensions of farm property lines, fishing grounds open to "dibs" each year, recognized usage rights which might follow property rights or deeds, and understood communal property where outsiders were not welcome. Stuart Beaton, in his work on the Northumberland Strait, claimed that the waters adjacent to farm property "were considered an extension of the boundaries of the farm" and "fence lines, hedgerows, etc. were extended out to sea as far as the 'hard bottom' or lobster habitat went." Every farmer/fisherman fished his own berth as he might farm his fields or cut his woodlot.[57] Similarly, focused in the area around St. Georges Bay, anthropologists John Wagner and Anthony Davis have made connections between recognized exclusive-use berths adjacent to farms and sociocultural fishing practices in eighteenth-century Scotland, where many local farmer/fishermen traced their origins.[58] In the homogenous community of Little River, Cape Breton, conflict in the late nineteenth century over non-owners hauling traps led the local Presbyterian minister to decree that lobster fishing was to be restricted to water adjacent to individual property. This system went unchallenged until the early 1980s.[59] While the formal law permitted open though regulated access, "lived law" made other claims.

For the first ten years after the introduction of new lobster regulations in 1873, local fisheries officers in New Brunswick continued to enforce what they referred to as "the practice of the country," where fishermen respected rights of prior occupation. Local officers enforced customary territory and settled disputes between lobster fishermen and packers who tried to fish in the same area. This ended in 1883. In August 1882 Shippegan Fisheries overseer Joseph Haché seized thirty-five traps belonging to John Young, a packer who he judged had crossed the understood line and encroached on the grounds of neighbouring canner Henry O'Leary. Young in turn sued the Department of Marine and Fisheries. Bureaucrats in distant Ottawa pointed out that the local practice had no basis in formal law. Officials on the ground attempted to preserve local practices, arguing unsuccessfully that they were supported by those involved. Until Young, fishermen and packers had respected the customary practice of "rights of prior occupation" and relied on the authority of the local state to resolve disputes. The formal legal opinion to the Department of Marine and Fisheries held, however, that "one fisherman has equal right with another" and no official had the right to seize traps.[60]

In many areas, especially in the Gulf of St. Lawrence and Northumberland Strait, cannery operators established implicit informal property rights over specific fishing grounds. This was also the case in Newfoundland, where historian Kurt Korneski notes that, in the late 1890s, local packers understood that when they purchased a factory, they also acquired exclusive rights to particular fishing grounds.[61] On the PEI side, packers often controlled 3.2–6.4 kilometres of shore, as this form of privatization stemmed in part from packers' practice of leasing boats and gear to local fishermen, who in turn were obligated to sell to them exclusively.[62] In these cases, claims of use rights were essentially transferred to the packers and not the fishermen who fished there. According to oral tradition, in Prince Edward Island packers' boundaries were "put in writing and 'set in verse,'" but as more fishermen were hired as "independent fishermen" rather than direct wage workers, the berths became "a 'gentlemen's agreement' among the packers." In the 1980s some lobster fishermen on the Island's North Shore claimed hereditary rights "due to the fact that their grandfather fished the same area fifty years ago for a small factory."[63] When large canning chains such as Burnham and Morrill moved into the Northumberland Strait area, they imported fishermen from the South Shore and later Eastern Shore of Nova Scotia and claimed local fishing rights.

The use of outsiders did not always sit well with local men. In the 1990s, one Northumberland fisherman recounted that when his father and uncle arrived at Pictou Island in 1923 from the Magdalen Islands and started fishing nobody seemed to mind. He added, "Try that today; it can get rough out there." Some fishermen believed that there was some formal law supporting the practice of exclusion. When fisherman Duney Duggan of Lismore, Pictou County, sold his boat and gear in 1937, he also signed a paper that carried moral if not legal weight in which he agreed "not to fish again on the said grounds of Lismore."[64]

Rights to specific fishing grounds, at least for the season, tended to be especially rigid in the Northumberland Strait where traps were attached to anchored long trawl lines and not moved during the season. During some periods, and some places such as the south side of Baie des Chaleurs in the 1890s, Prince Edward Island in the 1900s and the Northumberland Strait at least into the 1950s fishing grounds were annually up for grabs and fishermen raced out on the first day of the season to claim locations they believed to be the prime spot.[65] Fishermen annually competed to "stake their claim" for at least that season's location rights.

Defending these rights was another matter. Informal justice was not always employed or did not always work in successfully excluding outsiders. In these cases, some fishermen and packers turned to the state. The Department of Marine and Fisheries archives contains many examples of letters and petitions asking for help. As a fish dealer from Port Bickerton on Nova Scotia's Eastern Shore, H.S. Kaiser had a personal interest in local men having a successful season if he was to see returns on his investment. The dealer claimed he spoke on behalf of his neighbours for federal intervention as "it is the desire of the whole community to be protected against these intruders."[66] Dealers also petitioned to outside authorities at moments when the status quo was in flux. In 1905 Joseph Poirier, a canner in Grand Anse, New Brunswick, wrote to the Minister of Marine and Fisheries noting that packers in his district had "fished lobster quietly, each occupying his own grounds year after year, without being bothered by anyone," but in 1904 "a stranger came and commenced to surround the local fishermen's traps with his own." Poirier requested federal intervention to "stop this injustice." The letter was supported by a petition from other area packers demanding "that no one should be allowed to come and trespass, as it were, on these, what we consider to be our rights."[67] Even if informal law was impotent, these letters and petitions clearly articulated an understanding of justice and rights that was outside the formal law that any British subject could fish anywhere during an open season. As will be discussed in chapter 5, the most famous "lobster war" took place on the Northumberland Strait during the 1930s when a usage rights conflict resulted in arson, significant property destruction, and gunfire. In addition to these extra-legal strategies, local fishermen and packers successfully pressured the federal government to make it illegal in 1938 to fish more than one lobster district in any year.

Like the prohibitions on "outsiders" fishing in a local territory, the customary ban on Sunday fishing was eventually adopted into law. The custom against lobster fishing on Sunday seems to have been widely respected on Nova Scotia's Eastern Shore in the 1890s. Mondays meant poor fishing, as traps were not baited on Sunday and all lobsters that could escape did. Federal fishing statutes accommodated practices and laws and in 1914, opening days of the season that fell on Sundays were pushed back to Monday.[68] Similar accommodations were also made when the last day of the season fell on Sunday. The practice was still evident in 1957 when the National Film Board's "Trap Thief" suggested that Sunday was the prime time for lobster poaching in Cape Breton, as fishermen were in church and not on the water. The remains of this practice continue in southwestern Nova Scotia today where the season still opens on the last Monday of May.[69] The lobster statutes themselves did not make it illegal to haul a trap in Nova Scotia, New Brunswick, and Prince Edward Island on a Sunday until 1969, a ban that remained in place until it was legally challenged in 1984.[70]

Even when there was no conflation of formal and informal law, there are examples of fishermen writing to Ottawa in search of information, suggesting that at least some fishermen took considerable initiative to avoid fishing illegally. Bernard Smith of Port Medway, Nova Scotia, wrote in 1909 to "ask when the Lobster season opens as I can not find out here. Some says one thing and some another what I would like to know is does it start on or after 12 o clock on the 14 of Dec or at 6 o clock am on the 15th of Dec hoping to here from you before the 15 of Dec."[71] To Smith and the regulations these six hours mattered. The Fisheries Branch made ongoing efforts to educate fishermen about the laws. In 1894, an overseer on Nova Scotia's Eastern Shore called a meeting of lobster fishermen in every harbour to explain the newly introduced laws and regulations. This officer counted on his ongoing relationships with the fishermen and noted that as he knew "very many of the men in these harbours I thought I might influence them for good." The same year in the same area some thirty fishermen expressed different ideas about where regulations should come from and how they should be enforced. These fishermen came up with their own proposals and expressed the desire to select local fishery guardians themselves, proposals quickly rejected by Ottawa.[72]

Lobster fishing regulations were never evenly enforced, and their resoluteness was always contingent on place, time, and politics. Exceptionally, between 1888 and 1894, Minister of Fisheries Charles Hibbert Tupper made uniform and unpopular attempts to enforce the existing lobster fishery regulations and introduce new ones. This stringency waned with his successor, but he left the department on the path to more regulations.[73] The introduction of canning licences in 1894 coincided with attempts at policing canners, and in at least some districts the pre-existing leniency around the edges of the open and close season and minimum size disappeared. Local packers protested strict adherence to the close seasons, arguing that the regulations did not make sense of local conditions. Robert Holman, a prominent businessman and packer from Summerside, Prince Edward Island, wrote to his fellow Islander Minister of Marine and Fisheries Louis Henry Davies in 1896 that "the Department of Agriculture might, with as much reason, decree that all over the Dominion the grain harvest shall commence on a certain day as that your Department should insist that Lobsters shall be caught in all localities at one and the same time."[74] Pitting the freedom of the farmer in contrast to the restrictions on fishermen and canners was an argument employed by packers and politicians alike. The Liberal federal member for Guysborough when surveyed in 1902 as to whether there should be a limit on the number of traps and gear per fisherman replied no and that it would be the equivalent of restricting "how many acres of land the farmers of Manitoba should sow in wheat."[75]

In 1910 the bureaucrats in fisheries took the extraordinary measure of recommending to a special standing committee of the House of Commons on Marine and Fisheries regarding lobster fishery regulations that size restrictions be completely abolished, thereby abandoning a regulation always impossible to enforce. In exchange, the Fisheries Branch proposed that there would be more rigid protection of berried lobsters, more support for hatcheries, increased fees for canning licences, and an emphasis on conservation education for both canners and fishermen. In the district west of Halifax where existing size limits were the largest, all canning would be prohibited, and so size limits would be governed externally by the minimum import sizes legislated in Maine, Massachusetts, and New York where live lobsters were sold.[76] Canneries and their allied politicians pushed back on this last point, but size limits were generally abandoned until 1934.[77]

While size limits were rarely enforced, other regulations such as close seasons and catching berried lobsters were generally taken more seriously. The prohibition on catching berried lobsters had the greatest support of fishermen in many communities but was not supported everywhere. Prince Edward Island fishermen, notably Acadians, had a reputation for removing the eggs by striking spawn lobsters on the water to loosen the eggs and then removing them with their lobster mitts. In 1910 the Fishery Guardian at French Village, with the English name of James Feeham, argued that the majority of local fishermen supported liberating berried lobsters but the intense rivalry among them created a standoff, so they all removed the eggs.[78] The advent of motorboats provided a new technique for getting rid of eggs, as lobsters were placed under the exhaust pipe and throughout the 1930s allegedly almost every fisherman in Newfoundland used brushes to clean eggs off lobsters.[79]

Conclusion

The regulation of the lobster fishery was extremely complex, as the commons had a local context and was governed by both formal and informal regulations. The range of resource management strategies adopted can only be understood by considering both formal and informal approaches: the official policies that emerged were legacies of both streams. Formal laws and community custom coexisted and sustained each other in the same way that the marine commons and exclusive use co-evolved. Among the settler population, there was no widespread organized pre-capitalist lobster fishery. Individuals fished to supply the immediate needs of their households or communities. As a result, informal, community-based property rights, local management strategies, and vernacular forms of knowledge co-evolved, with a new Canadian state trying to establish its own formal, legal regulations, and industrial production. In addition, starting in the ecological crisis of the 1880s when the lobster stock collapsed for the first time, the Canadian state had to resist (and in at least one case actively cooperated with) canners who attempted to establish property rights over specific areas of coast and water. By the late 1960s fishermen's pragmatic use of the state to protect local access had ultimately undermined more general claims of rights based on a shared marine commons, and it was a turning point in the path to a limited access fishery. Fishermen, packers, and the state rarely shared the same interests. Biological, economic, political, and community forces combined to create a regulatory regime that was always contested and perpetually in flux as each party sought to reap their own social, economic, and political rewards.

Chapter Four

The Politics of Knowledge: The Lobster Question, 1874–1939

Hon. Mr. BRODEUR. – Let us inquire into the question of lobsters.

The CHAIRMAN. – We are going to take up the lobster question first because it is said to be the most important, the most pressing question in the maritime provinces at present.

Evidence taken before the Marine and Fisheries Committee respecting the Lobster Industry during the Session of 1909, p. 4

At the core of the state's regulation of the lobster fishery was knowledge. Information, evidence, and science were central to how the state positioned itself as a neutral and objective mediator, arbitrating between competing interests and justifying its policies as rational and fair. Yet knowledge was never truly neutral. What counted as information was created, collected, and interpreted within particular social and political contexts; it could be privileged or ignored, accepted as truth or dismissed as misrepresentation. In the case of lobster, knowledge emerged both in local ecological understanding and the electoral calculations and administrative frameworks of Ottawa-based federalism. The stakes were high: by the end of the nineteenth century, the commercial lobster fishery was Canada's second most important Atlantic fishery (and in some reporting years, such as 1898–99, it surpassed the value of groundfish) For many Maritime coastal communities it was essential for survival.[1] The precise lobster question alluded to at the opening of this chapter was never explicitly articulated. If it was, I would imagine it to be – what government policy would best protect the resource yet maximize immediate profits and political gain for the party in power? Of course, there was no magical answer to these irreconcilable objectives, so various answers were refracted through prisms of the divergent self-interest of fishermen, packers, local employees in the Department of Marine and Fisheries (and later Department of Fisheries), local and national politicians, bureaucrats, and scientists.

This chapter focuses on the period 1873 to the Second World War when fisheries knowledge was characterized by both a local practical and empirical orientation. Information was often based on counting and anecdotes; what experts, fishermen, and government employees thought they saw in specific places. This meant the figures connected to the number of cans, cases, traps, and men employed created detailed tables illustrating change over time, organized chronologically to reveal positive and disastrous trends. Some people were completely absent from all discussion. There was never even a suggestion of Indigenous concerns. The perceived challenge became conveying very local insight about lobsters to Ottawa and accommodating this specific ecological knowledge within an Ottawa-centred agenda that privileged the place-based dynamics of electoral politics and was attuned to the broader complexities of the federal system. Commissions and inquiries were a strategy to bridge the chasm between local knowledge and federal politics as they highlighted local expertise, made policy recommendations, and dissipated political discontent. They also broadened the range of interest groups consulted. This effort to connect local and federal agendas contrasted with the postwar period when the nature of knowledge and expertise changed and became more abstract and specialized. It moved from pragmatic observation to models and theories proposed by economists and mathematicians. Pre-Second World War science was more practical, empirical, and intensely political, shaped primarily by economic understandings of resource extraction. If scientists were predisposed to see the lobster industry from an economic point of view – it is not without irony to note that subsequently their influence was surpassed by economists.

There is a generalization in the history of fisheries that has contrasted the place-based, traditional, or experience-based knowledge held in local communities against the scientific knowledge of outside experts.[2] This was never so simple for the lobster fishery. The example of lobsters was parallel to historian Christine Keiner's observations about the Chesapeake oyster fishery where a combination of traditional and scientific advice (until the scientists sided with private interests) meant that community practices emerged concurrently with scientific opinions.[3] Outside experts were present from the start; again, large-scale exploitation of oysters was not a traditional fishery that had established historical community-based practices.

The management of fisheries under the new Dominion government focused on information gathering. Data was collected, especially through counting at the local level to report on the health of fish stocks and make claims in international treaty negotiations with the Americans. Lobsters did not fall into these international negotiations (which scholars have noted were focused on establishing national sovereignty) and therefore were slow to appear in official reports. Fishery science was limited in the nineteenth century, and fisheries officials depended heavily on transnational networks of expertise, which in the case of lobster meant Americans. But knowledge was also generated in local communities – by fishermen and by packers who made claims based on established practices, and through many investigative government reports, inquiries, and commissions. In 1896 Commissioner of Fisheries Edward Ernest Prince (1858–1936) proposed that as the "fishermen and canners wholly disagree on the most vital points," scientific investigation could somehow bridge this gap.[4] A language of objectivity associated with science attempted to "depoliticize" the regulation of the industry, although the management was always deeply political.

Before the First World War, the federal Department of Marine and Fisheries was an important and large department by measures of spending and in terms of political influence. In 1905, for example, only the Departments of Public Works, Railways and Canals, and Finance had larger departmental budgets.[5] Between 1892 and 1930, the Fisheries Branch of the Department of Marine and Fisheries did not have its own distinct deputy minister but was led by a Commissioner of Fisheries (and General Inspector of Fisheries for Canada) with daily operations under the direction of the Assistant Commissioner of Fisheries (renamed the Superintendent of Fisheries in 1903). There was remarkable stability in the Fisheries Branch's leadership. Prince served as Commissioner of Fisheries between 1892 until his retirement in 1924.[6] His deputy was Robert Norris Venning, a man who was especially well suited to the complex worlds of Ottawa and fisheries. His father, William H. Venning, had been the Inspector of Fisheries for Nova Scotia and New Brunswick at the time of Confederation, and the younger Venning worked in the Department of Marine and Fisheries in Ottawa from 1873 until his retirement in 1911.[7] Venning was replaced by Prince Edward Islander William Ambrose Found, whose position was renamed Assistant Deputy Minister and recast after Prince's retirement as Director of Fisheries. Once the Fisheries Branch became a separate department, Found served as Deputy Minister from 1930 until 1938. Altogether, for a remarkable twenty-seven years, Found ran federal fisheries. Although both Venning and Found had roots in the Atlantic region, they were firmly Ottawa-based, and it was a regular complaint of politicians that those responsible for fisheries policy did not spend enough time in the Maritimes.[8] Spending time in communities was another way of knowing.

The Department of Marine and Fisheries predated the emergence of fisheries biology as a specialized field. Historian Jennifer Hubbard has argued that fisheries biology helped "define territorial waters and protect and extend state interests" and emerged from "local and particular political and economic interests." The same observations hold true for the Department of Marine and Fisheries. Both the evolving science and the department perpetually negotiated the contradiction, particularly obvious in the example of lobster, between conserving the fishery while at the same time expanding it for economic development and prosperity. Lobster differed from other sea fisheries. The foundational premise for all lobster regulations was the conservation of "Canadian" stocks after the Americans had purportedly destroyed their own resources through overfishing.[9] The immediate suspicion of US canners was no coincidence, as this was the exact moment that the Washington Treaty of 1871 came into effect. Canadians have generally understood this treaty as an example of Britain sacrificing Canadian interests to appease Americans – especially in granting Americans wide access to the Canadian fisheries.[10]

The early employees of the federal department who shaped lobster regulations were at best talented, self-taught, practical men or at worst incompetent patronage appointments. In the 1870s and 1880s they were exclusively men of "practical experience" who liked to count, measure, and weigh and were closely connected to their international counterparts, especially in the United States. They had little or no formal biological knowledge, but their preoccupation with numbers proved extremely useful in international negotiations with Americans who, with fisheries falling under state jurisdiction, lacked comparable centralized statistics. Collecting statistics was at the core of the Department of Marine and Fisheries' work. At an international fisheries conference in London, England, in 1883 Brown Goode, United States Commissioner of Fisheries, stated that the Canadian department "was one of the most valuable organizations in the world," as its collection and publication of statistics was an international model. While the department was aware that its numbers were incomplete and inaccurate, it insisted that they were still useful in providing trends. The "decay" of fish stocks was receiving international attention, and in Britain concerns about declining fish stocks prompted the government to respond by establishing investigatory commissions.[11]

One of the veterans of Thomas Huxley's 1884 British commission, who had done graduate work at Edinburgh and Cambridge, left his post as

Professor of Zoology and Comparative Anatomy at Glasgow's St. Mungo's College in 1892 to become the Dominion Commissioner of Fisheries. Professor Prince was the first civil servant to lead fisheries with a formal scientific background. Sociologist John Flint described Prince as a "Victorian conservationist" who applied scientific methods to resource management to achieve economic ends.[12] With the arrival of Prince and the development of fish science, the department developed relationships with Canadian academics who were issued contracts to conduct specific research projects or inquiries. In his first report as Dominion Commissioner of Fisheries, Prince noted that of all fisheries, the least "reliable information" was available about lobster and recommended that specific research on lobsters' "habits, propagation, life-history and migrations" could be the basis of "wise legislation."[13] In 1894 and 1896 Prince himself authored and published an "Inquiry into the Lobster Fishery" and a "Special Report of the Natural History of the Lobster, with Special Reference to the Canadian Lobster Industry," respectively.

Another generalist who had influence before the First World War was William Wakeham. Wakeham was the officer in charge of the Gulf fisheries and had a general science background as a McGill University trained physician. From 1879 until 1909, representing the older generation of hands-on non-specialists, he performed his diverse fisheries duties and served on commissions and Arctic expeditions.[14] This practical orientation and knowledge acquired through lived experience meant that his cohort – while patronizing and condescending –acknowledged the expertise of fishermen. Academic scientific authority was ascending, but not in complete control. The introduction to the 1898 Royal Commission on the Lobster Industry pointed out that fishermen possessed a large amount of real information that they had willingly shared with the Commission. Prince who wrote the report expressed surprise at their "power of observation ... considering the opportunities which most of them have."[15] The same attitude is mirrored by Wakeham in his 1910 investigation into the lobster fishery. In this inquiry, he appeared to have followed a basic methodology of interviewing the oldest fisherman he could find in any given harbour and asked him how long he had been fishing lobster and what had changed. He concluded his own report emphasizing the authority of his own experience. Wakeham's investigation had been the "honest effort of one who has had over forty years connection with the fishing coast and its people, to suggest some practical and plain regulations for the preservation of the fishery, which should be practical to enforce, with a minimum of friction and cost."[16] This pragmatic approach based on direct involvement in the area was being eclipsed by academic specialists.

The Fisheries Branch recruited outside experts, usually academics, for their scientific expertise. In 1898 Bishop's Medical Faculty's Dr. Andrew Macphail was engaged to investigate why canned lobster often turned black, which harmed the reputation of Canadian lobsters, as it rendered the product unsaleable. Here, a university position, medical degree, and Prince Edward Island origins provided sufficient credentials. The academic most closely associated with lobster in the Fisheries Branch before the 1920s was Archibald Patterson Knight (1849–1935) of Queen's University.[17] Knight was Professor of Biology and Physiology, which meant that he was both an expert on lobster hatcheries, conservation, and canning and the author of the 1910 standard Ontario school textbook on human hygiene. Knight was research director of the Biological Research Board from 1912 to 1922 and focused on pragmatic issues such as lobster hatcheries and strategies to raise the standards of canning. His broad general expertise did not lessen his arrogance as he introduced his talks to local fishermen by saying, "You'd better listen: I know more about lobster than any man alive."[18] Claims of specialized scientific knowledge could be tenuous. Knight's investigation into lobster pounds conducted in 1915, but not published until after the war, argued against the use of pounds for propagation as "all beings suffer from confinement – humans in jails, asylums and detention camps – no wild animal flourishes so well in confinement as in the open."[19] The conflation of the human and lobster experience would not be adopted by later, more specialized scientists.

In 1898, after lobbying by E.E. Prince and others, Parliament began to fund a floating biological research station on the East Coast. This brought outside experts, at least seasonally, to the local context. That year, the Royal Commission on the Lobster Industry had to rely on British and American research, as little work, except for Prince's publications, had been done in Canada. The use of the new station led the government in 1908 to open two permanent coastal centres, the Atlantic one located in St. Andrews, New Brunswick, selected in large part because of its direct rail link to Central Canada.[20] After 1912, its management came under the Biological Board of Canada, but there were no permanent employees until 1925. Ambivalence remained within the department towards the St. Andrews station's pure research orientation and after the department refused to purchase German and French language scientific publications, the Board lobbied Parliament for its own budget. In its first years, the St. Andrews station served as a research facility for academics and their students who moved scientific knowledge towards more problem-centred studies rather than making an inventory of resources. This group tended to be general biologists who were eclipsed in the 1930s by marine specialists and the Fisheries Research Board.[21]

The most important scientific project of the turn of the twentieth century was one that fishermen, canners, and politicians could all support. The artificial propagation of lobster eggs through government hatcheries was a politically expedient solution to conserve the species and support stock management without restricting anyone's actions. Lobster hatcheries had the additional benefit of creating opportunities for local partisan spending and the distribution of patronage positions. It was a win-win-win situation with more lobsters for fishermen to catch, more lobsters for packers to can, and none of the political backlash associated with restricting catches. The science behind fish hatcheries in the 1860s expanded possibilities. In the United States, where the federal government had no local authority over fishery regulations, hatcheries were a practical solution for conservation at the national level, as this was a legitimate federal action. For lobster, Canada took direct inspiration from a hatchery operating in Dildo Island, Newfoundland, using methods developed in Norway.[22]

Lobster propagation was one of the initial programs of the Atlantic Biological Station. In 1908 there were already five government hatcheries in operation, located for political benefit as much as ecological context.[23] In addition to operating government hatcheries, the Department of Marine and Fisheries entered a private arrangement in 1904 with canner H.E. Baker, who was affiliated with Roberts, Simpson and Company. Under the agreement, Baker was to be permitted to impound live seed lobsters at Gabarus, Nova Scotia, and release them at the end of the season. He was paid 16¼¢ per lobster to a maximum of 50,000 – an arrangement that yielded a tidy profit for Baker.[24] The actual benefit to lobster stocks, however, was dubious. What the policy did achieve was political. It responded to the grievance of many Maritime fishermen that "Canadian" berried lobsters (female lobsters carrying eggs) ended up in American hands to the benefit of the US fisheries.[25]

The numbers circulated were impressive, if largely meaningless. In 1914, for example, eight government-run lobster hatcheries claimed to have released 700 million lobsters.[26] The first hatchery opened near Pictou at Bay View in 1890. Between 1903 and 1913, another thirteen hatcheries were established, each claiming to produce at least 50 million young lobsters annually. From 1903 onward, most of these operations were built and run by the Liberal government.[27] The Bay View hatchery itself was constructed under a Conservative administration in a Conservative riding. It was operated by Alfred Ogden, a former lobster canner in Canso. His more significant credential, however, was

Figure 14. Postcard Lobster Hatchery, Charlottetown, P.E.I. (Toronto: Rumsey and Co., n.d.) Courtesy of the Robertson Library, University of Prince Edward Island

political: Ogden had served as a Conservative Member of Parliament for Guysborough from 1878 to 1882. Although he was defeated in subsequent elections, he continued to benefit from federal patronage.[28]

The locations of the hatcheries were unabashedly political and not surprisingly they were almost always located in a federal riding that had elected the party in power. All construction spending went through the Minister, but local Members of Parliament influenced decisions such as the location and who was hired. The Liberal Member of Parliament for Westmorland, for example, cautioned against the establishment of a hatchery at Cape Tormentine, as residents to the east "could be very much dissatisfied if the hatchery was placed there," as it was not equidistant between the district's two large canning areas.

Expertise in hatching lobster eggs was ignored. The Prince Edward Island Member of Parliament for East Queens lobbied for the establishment of a hatchery inside Charlottetown Harbour, although the location was completely unsuitable because of pollution and fresh water. The compromise was a hatchery at the west entrance of the harbour at Blockhouse Point, but it proved to be a constant source of aggravation with ongoing ice and silt issues. Fisheries bureaucrats' recommendation

that it be closed and relocated was vetoed through political intervention. Spending and contracts also followed political loyalty. At Canso,
A.N. Whitman and Son sold the property for the lobster hatchery
and received the contract for the collection of local lobster eggs on its
steamer. Both contracts were arranged through John Howard Sinclair,
Liberal Member of Parliament for Guysborough.[29]

With some exceptions, politics not competency generally dictated
employment. The position of hatchery "officer in charge" was a plum
patronage job, as it received the salary of $3 a day in 1903.[30] Westmor-
land Liberal Member of Parliament from 1900 to 1914, Henry Robert
Emmerson involved himself in every aspect of the local hatchery while
his party remained in power. In December 1904 he wrote to the Minister of Marine and Fisheries asking that two men not be reemployed
during the 1905 season and gave the names of specific replacements.
At the same time, he lobbied successfully that the chief engineer from
a disastrous 1904 season be re-engaged.[31] In December 1907 Emmerson
was again lobbying a new Minister of Marine and Fisheries, stating
that as a result of complaints from packers, he was personally recommending all the employees to be hired in 1908. The Shemogue Lobster Hatchery from 1904 was headed by Napoléon LeBlanc, who had
worked "in the interests of the Liberal party." Throughout his tenure,
there were regular complaints and reprimands related to poor management and supervision. Supplies were not ordered, and he was paid for
days when he was home making hay and the hatchery was closed.[32]
The defeat of the Liberal government in 1911 meant that LeBlanc could
finally be replaced.

In northern New Brunswick, the Shippegan hatchery required the
local Liberal Member of Parliament's approval for all employment and
community spending. Names of prospective employees were provided
by a committee of the Member's "friends" in Shippegan.[33] After the
Conservatives were elected nationally in 1911, the local party candidate
Theobald M. Burns wrote to the new Minister of Marine and Fisheries
to explain that the hatchery staff had been active during the campaign
leading to his defeat. He complained that the manager and employees had "near precipitated a row and bloodshed" when they tried to
break up a Conservative meeting at Ste.-Marie Lemèque.[34] The new
Minister laid out his dilemma as although the staff was clearly partisan, the Shippegan hatchery was a success, with both a good record
and the lowest expenses of all hatcheries.[35] Burns persisted to plead his
case that the current employees were "actively engaged in canvassing
and working in the interest of the Liberal Party."[36] As a compromise,
the Minister asked that the manager who had been in place since 1903

be retained but that the defeated Conservative candidate could name the rest of the staff. The manager ended up resigning in July 1912 after threats of personal violence, and the Shippegan hatchery was closed in 1914 because of ongoing political conflicts.[37]

In the end, A.P. Knight, chair of the Biological Board of Canada, used his own research to shut down the lobster hatcheries following the First World War.[38] In a 1918 announcement that all fourteen lobster hatcheries would close, the *Canadian Fisherman* reported on Knight's research that showed only 100,000 of the 62 million eggs at the Georgetown, Prince Edward Island, hatchery in 1917 were viable at a cost of $2,500. Knight argued that this low productivity did not justify the efforts and expense and the removal of eggs from the natural environment might even contribute to stock decline. The answer, according to Knight, was another form of knowledge that placed most of the responsibility elsewhere – educating fishermen on conservation.[39]

The lobster egg hatcheries bent science for a politically expedient strategy to address declining stocks without limiting the liberty of either packers or fishermen. For over twenty years, results were ignored or misread. This use (or misuse) of science for economic and political goals shared things in common with the department's nearly hundred-year-long effort to introduce lobsters into the Pacific Ocean.

Pacific Lobsters?

The Ottawa-based Department of Marine and Fisheries had to contend not only with local Maritime concerns but also the geography of Canadian federalism. The dynamics of regional resentment came into play as there was an imbalance between the substantial fisheries revenue that British Columbia's salmon fishery generated and the sparse amount the federal government invested in fisheries in the region. From the 1890s until after the Second World War, the British Columbia salmon fishery dominated Canadian fish exports. Instead, the spending in the Federal Department of Marine and Fisheries reflected the allocation of electoral seats and the number of workers employed in the industry, not the revenue generated. The department responded with various failed attempts to introduce lobster into the Pacific Ocean. Starting in 1887 and until 1966, the Canadian government made at least eight unsuccessful attempts to release lobsters off the coast of British Columbia. Three additional efforts were made by, or in conjunction with, the private sector. The project of introducing lobster provided a politically expedient strategy for addressing British Columbia's political concerns, if not a happy tale for the lobsters.

The scheme to create a new and valuable fishery on the Pacific was ostensibly an economic project to provide "an additional source of wealth to that portion of the Dominion."[40] Economic motivations were the most common justification for intentional species introduction. Nature was a commodity that could generate wealth, self-sufficiency, or at least human enjoyment. The repeated attempts to introduce Atlantic lobsters into the Pacific Ocean, however, were also rooted in politics. Here again political interference dictated government fishery science.

Americans identified the economic possibility of introducing Atlantic lobsters to the Pacific Ocean in the 1870s once their transcontinental railway was completed. Attempts by the US Fish Commission and the California State Fish Commission resulted in dead lobsters and the occasional "feeble" live lobster reaching San Francisco Bay. By the late 1880s Northern California, Oregon, and Washington were identified as having the same ocean temperature and salinity as the East Coast and efforts were redirected northward.[41] American efforts intensified again in the 1906–17 period, with a focus on Puget Sound and Yaquina Bay, Oregon. This coincided with the height of Canadian lobster egg hatchery expansion off the East Coast.[42]

Pressure on the Canadian Department of Marine and Fisheries to act came in part because of American efforts, but the abject failure of American attempts had no effect on Canadian ambitions. Exactly two years after the completion of the Canadian Pacific Railway, the Department of Marine and Fisheries made its first attempt to transport live lobsters for breeding purposes to British Columbia. In November 1887, live lobsters were packed in Dalhousie, New Brunswick (the closest railway connection to Vancouver with lobsters but over 5,300 kilometres away), but they all died in a cold snap. A second attempt was made the following spring.[43]

The main proponents of this scheme were British Columbian politicians. Conservative Edward Gawler Prior, who represented the City of Victoria from 1888 to 1901, was joined in 1890 by fellow Conservative Gordon Edward Corbould representing New Westminster. Encouraged by American experiments, Prior and Corbould lobbied the Department of Marine and Fisheries, going so far as to claim in the House of Commons that three American lobsters had immigrated into British Columbian waters. The proactive Minister of Fisheries Charles Hibbert Tupper (who was ultimately a much more successful Nova Scotia transplant to British Columbia than any lobster) reported that his department was watching the Americans but wanted to avoid unnecessary expense until the scheme proved to be feasible.[44]

Once Tupper left the portfolio, politics trumped evidence as Conservative fortunes in British Columbia came under assault. In July 1894, Fishery Overseer Alfred Ogden was instructed to send lobster eggs to a hatchery in New Westminster, British Columbia. The Minister, conscious of the need for local electoral success, as the Conservatives held all five British Columbia seats, noted that this project was "of considerable importance," as it had the potential to bring additional prosperity to the region.[45] It was determined that April and May were the ideal months in terms of temperatures and that a newly available (but very primitive) refrigeration car could keep lobster eggs alive if they were packed in wet sea moss.[46] Even though sending eggs was much less expensive than sending actual lobsters, this program was halted by departmental budget cuts in 1895. This decision was met with further lobbying by a British Columbia Member of Parliament who failed to see any reason "the lobster fishery should not assume gigantic proportions on the Pacific, as well as on the Atlantic coast."[47]

The Fisheries Commissioner in 1896 boasted that Canadians had learned from the "mistakes" of Americans in their "costly and useless" attempts of the 1870s and 1880s. It is difficult to see how, when the same year the Department arranged a special baggage car (not refrigerated!) to ship 600 live lobsters west in the heat of July.[48] The timing of this shipment was strange (not only because it was summer, and lobsters really need to be kept cool) but as the shipment arrived immediately after the 1896 General Election that brought the Liberals to power.

From a purely scientific perspective, it seemed curious that the university-trained scientist and specialist in fish culture, Fisheries Commissioner Prince with no specific evidence beyond the claim that 500 lobsters had arrived alive, concluded that the experiment was a great success. The enterprise was repeated the following year and lobsters were planted in four different locations off the mainland and Vancouver Island.[49] In 1898 Prince reported that the lobsters "must have bred by this time" and that the introduction of lobsters should be repeated, but in May, before the warm weather made keeping lobsters alive on a transcontinental journey difficult. The enthusiasm for the project, considering the lack of any scientific evidence of its success, was made explicit by Prince's observation that "it would be very popular and would do much to allay the dissatisfaction resulting from the allegations that the Department derives very large revenue from the British Columbia fisheries and has little in return."[50]

In 1905 a fourth attempt was made, and lobsters were released in six locations on the Lower Mainland and off Vancouver and the Gulf Islands. Ultimately this trip was more about the success of transportation advances. Live berried females were crated and wrapped in rockweed packed with ice that was replenished more than fourteen times over the journey from Halifax to Vancouver in five and a half days.[51] Their dispersion revealed the politics of the enterprise. British Columbia Members of Parliament selected the release sites – not civil servants with any specialized knowledge.

Considering the political stakes for the department, explanations needed to be found for why a Pacific lobster fishery was not being established. The failures followed established national narratives. Much like concerns about Canadian outmigration to the United States, in 1901 the Minister of Marine and Fisheries concluded that strong ocean currents were sweeping the young lobster fry into American territory. "Hence it has not been possible to establish any clear and unquestionable benefit to our waters."[52] When the 1905 shipment of lobsters arrived in British Columbia, they were accompanied by specific instructions from the Commissioner of Fisheries that the lobsters in Nanoose Bay needed to be protected from being poached by Indigenous fishermen. This concern reflected dominant tropes about Indigenous overhunting.[53] Reaching beyond these tried-and-true excuses, the Dominion superintendent of fish culture in 1907 insisted that "non-capture does not demonstrate that attempts have proven failure." He continued to emphasize that "the practical and common sense side of fish breeding" was as important as scientific research and recommended that Canadian Fisheries Museum curator Andrew Halkett, who was already in the West, visit the coast and apply his familiarity with East Coast lobster pounds to see if conditions might warrant another attempt.[54] So twenty years into the project, practical rather than political issues were finally addressed (but not by anyone with real expertise in aquaculture) and the department's official position was that the introduction of lobsters into the Pacific could be judged as neither a success nor failure.[55]

And so, it continued. In April 1908 (the same year that Atlantic oysters were also introduced into the Pacific Ocean) lobsters and lobster eggs were shipped from Halifax in six days, with 1,100 lobsters making it alive to Sooke Harbour. The Dominion Commissioner had expressed opposition to the use of Sooke Harbour as a release site since it was too close to the United States and had a muddy bottom, but he was overruled.[56] The following year 1,750 lobsters were "liberated" off Mudge Island near Nanaimo. Many of these lobsters were kept in pounds for five weeks until a storm broke their netting, and this successful short-term captivity was interpreted as evidence that an Atlantic lobster could survive in Pacific conditions.[57]

When planning for the 1909 season, unsuccessful attempts were made to catch any surviving Pacific lobsters. Now Indigenous people were useful in supporting this enterprise, as local fisheries Inspector E.G. Taylor reported that the "Indians at Sooke stated that they had seen lobster last fall on more than one occasion." This rumour was repeated the following year.[58] During the winter of 1910, the BC Fisheries Inspector reported that "two White Fishermen," described as "reliable" witnesses, presumably in contrast to the Indigenous testimony, had seen lobsters. Further investigation concluded that any lobsters found were those that had broken free from holding pens or were large sand shrimp.[59] One wonders how attempts to catch lobsters were being made. A memo by the Fisheries Commissioner in 1912 noted that there were no lobster traps in the Pacific region and "I do not think the crab pot in use there, would be successful for lobsters." Subsequently, two dozen wooden lobster traps were shipped west.[60]

As Americans embarked on another large introduction of lobsters into the Puget Sound in 1913, Canadian scientists associated with the Biological Board of Canada lobbied the Marine and Fisheries Deputy Minister to follow suit.[61] It is significant that the push for continuing introduction of lobster came not only from British Columbian Members of Parliament and Boards of Trade but from professional scientists who hoped success would increase their influence. In 1914 Professor Archibald B. Macallum, of the University of Toronto's Medical faculty, one of Canada's leading scientists in medical and biological research and later founder of Canada's National Research Council, enthusiastically supported the program as a member of the Biological Board of Canada. Again with no empirical evidence, he stated that as conditions were similar, "there seems no reason why they should not do well ... Personally I should not be surprised if they have hunted up some locality that suited them well and are still alive and thriving increasing in number."[62]

Although evidenced-based decision-making eventually led the Department of Marine and Fisheries to close its considerable network of lobster hatcheries after the First World War, evidence was less of a factor for the introduction of lobsters to the Pacific. Indeed, in 1946, in a public/private partnership, lobsters from Prince Edward Island were brought to the Pacific Biological Station, Departure Bay, Nanaimo by British Columbia Packers Limited. Half of the lobsters were confined at the station (and died within two days), while 1,605 lobsters were released into a nearby lagoon. Three lobsters survived until 1948. Notwithstanding this failure, a November 1947 industry report described the experiment as "going well" and kept the location secret to dissuade poachers. In 1949 the industry press continued to claim inaccurately that released lobsters survived, although it admitted they did not breed in Pacific waters. Completely private efforts were made in 1954 and 1956 when lobsters from St. Andrews were transplanted to Prince Rupert. Finally (so far) in 1965 and 1966, the Department made the last attempt at lobster introduction, when it imported lobsters from Prince Edward Island to British Columbia again.[63] Those behind this effort published *Lobster (Homarus americanus) Production in British Columbia* in 1972 and again concluded (this time using the modelling of marine ecosystems fashionable at that time) that a "commercial lobster population is feasible at selected sites on the Canadian Pacific coast" and that "a colony of lobsters could be established."[64]

The long history of failed attempts to introduce lobsters into the Pacific Ocean, like the history of lobster hatcheries, reveal the challenges for fisheries bureaucrats practising science in a political context. In both cases, the easy and popular solutions to expand a vulnerable stock or create a new resource in an entirely new ecological context did not work. Fledging government scientists had their authority contested, not only by politicians but also by fishermen and packers who hoped knowledge and information would serve their own distinct objectives.

Scientific expertise could be dismissed with disdain. In 1918 Moses H. Nickerson, who claimed authority based on fifty years in the lobster business, a Liberal political career largely related to the fishery, and service on the 1898 Royal Commission on the Lobster Industry rejected government claims of a lobster stock crisis. "The Government has given a lot of college dons the ultimate say on lobsters!" he raged. "Who is Professor Knight?"[65] The same year, an article in the *Canadian Fisherman*, written by an unidentified canner, claimed that the scientists were "prone to accept unreliable reports as fact." Government statistics were incorrect and exaggerated and any decline in the lobster stock was not serious.[66]

Commissions

The critic Moses H. Nickerson had a direct voice in shaping lobster policy through his participation in a government commission. Commissions and inquiries served as an important source of gathering local evidence and popular opinion through a supposedly objective and fact-finding forum. They created space to generate another kind of knowledge, information, and understanding of the lobster problem that was experience-based and locally situated by those directly involved such as Nickerson. But all investigations, inquiries, and commissions had concurrent information gathering and political goals. And indeed, no commission could reconcile the difference of how "facts" were interpreted by different interest groups. In 1899, for example, while most of the members of the Royal Commission concluded that the condition of the lobster stock was "approaching 'critical' or already critical in some localities," two commissioners closely associated with packers dissented from the final report. They maintained that there was no problem, as "natural conditions prevented excessive drain."[67]

Between 1887 and 1928, the Canadian federal government conducted fifteen special committees, inquiries, or full royal commissions, either directly concerned with or touching upon the state of the lobster industry (see Figure 15). These forums in turn generated their own kind of knowledge. The composition of these inquiries was contentious. It had to reflect industry and other political stakeholders through economic, geographic, religious, and linguistic diversity and accommodate claims of expertise as defined by practical experience in fishing, canning, or more abstract research or academic credentials. Although no active lobsterman ever served as a commissioner, these investigations appear to have become more inclusive or democratic over time, holding hearings in disparate fishing and canning communities. The 1927 Royal Commission on the Fisheries of the Maritime Provinces and Magdalen Islands facilitated fishermen's direct participation as witnesses through reimbursing the transportation and per diem expenses of two delegates from every harbour.

At a time when research capacity was limited, commissions, inquiries, and special reports played an important role in shaping what the Fisheries Branch thought it knew. In the words of sociologist Adam Ashforth, commissions provide a "common language through which to speak of the problems and articulate solutions." E.A. Heaman, in her work on taxation, noted that the measure of a crisis can be read in the proliferation of related commissions.[68] With this measure, the lobster industry was a burning issue.

The crisis in this case was not only a perceived ecological collapse of lobster, but its political, economic, and social consequences. Canadian law established two different kinds of commissions. Major commissions usually labelled Royal were appointed by Order in Council and took their authority directly from the Crown. Departmental investigations, on the other hand, were with or without reference to the Inquiries Act and proceeded on behalf of the Department.[69] During the period 1867–1929, different prime ministers used commissions differently. Historian Patrice Dutil has pointed out that in the 1880s, John A. Macdonald increased his use of commissions especially in the areas of social and

Table 1. Commissions and Inquiries, 1874–1927

Commission, Inquiry or Special Report	Author or Commissioners
1874 Special Report on Certain Petitions Against the Regulation Affecting the Lobster Industry	W.H. Venning, Inspector of Fisheries, NS & NB
1887 Lobster and Oyster Fisheries Commission	Edward Hackett, Tignish, PEI Alfred Ogden, Halifax, NS Wm. B. Deacon, Shediac, NB J. Hunter-Duvar, Alberton, PEI, Secretary
1889 Report on Lobster	Lt. A.R. Gordon, Fisheries Protective Service
1892 Special Report on the Lobster Industry	William Smith, Deputy Minister of Marine and Fisheries
1894 Inquiry into the Lobster Fishery	Edward E. Prince
1896 Special Report on the Natural History of the Lobster, with Special Reference to the Canadian Lobster Industry	Edward E. Prince
1898 Royal Commission on the Lobster Industry	Edward E. Prince, Chair Moses H. Nickerson, Clark's Harbour, NS Wm. Whitman, Guysborough, NS Donald Campbell, Margaree Forks, NS Henry C. Le Vatte, Louisbourg, NS Archibald Currie, Souris, PEI Stephen E. Gallant, Egmont Bay, PEI Patrick J. Sweeney, Shediac, NB Robert Lindsay, Gaspé, QC
1903 Commission to inquire into the Herring and Sardine industry in the Bay of Fundy, as well as the ravages of the Dog-Fish and the General Condition of the Lobster Fishery at the Magdalen Islands, St. Mary's Bay, and the Bay of Fundy	Joseph J. Tucker, Chair, Saint John, NB Robert Norris Venning, Ottawa Rev. Joseph Samuel Turbide, Havre-aux-Maisons, QC Albert James Smith Coop, Digby, NS E.C. Bowers, Westport, NS Robert Edwin Armstrong, St. Andrews, NB
1904 Gloucester Dogfish and Lobster Commission	Peter (Pierre) Morais, Caraquet, NB
1909 Comprehensive Study of the Lobster Industry	Select Standing Committee, Ottawa
1910 Commission on the Lobster Industry in Quebec and the Maritime Provinces	Wm. Wakeham, Officer in Charge of the Gulf Fisheries, Gaspé
1912 Commission on Shellfish	Edward E. Prince, Chair Richard O'Leary, Richibucto, NB John McLean, Souris, PEI S.Y. Wilson, Halifax, NS

	Commission, Inquiry or Special Report	Author or Commissioners
1915	Report upon Lobster Investigation at Long Beach Pond, NS	Professor A.P. Knight, Kingston, ON
1919	Commission to Investigate Illegal Lobster Fishing in the Shediac Sub-Division of New Brunswick	Ward Fisher, Assistant to the Superintendent of Fisheries
1927	Royal Commission on the Fisheries of the Maritime Provinces and the Magdalen Islands	Alexander K. Maclean, Chair, Halifax, NS Cyrus Macmillan, Montreal, QC Henry Ryder Locke Bill, Lockeport, NS Joseph George Mombourquette, L'Ardoise, NS John George Robichaud, Shippagan, NB G. Fred Pearson, Halifax, NS, Counsel

economic issues. Dutil concludes, however, that although Macdonald was "interested" in these commissions' findings, he did not necessarily adopt their recommendations but used their conclusions as a strategy to introduce new ideas to the public. Wilfrid Laurier turned to commissions even more than Macdonald, and Robert Borden adopted their use even further.[70] In establishing a commission, politicians sought to assert empirical-based, unbiased decision-making through different forms of authority and supposedly "depoliticize" a very political process. The state could fashion itself as a neutral actor, its objectivity buttressed by evidence. So, commissions, inquiries, and reports pacified, educated, and sometimes prepared the way for controversial policies.[71]

The rules of evidence in a commission meant that while witnesses were sworn in before they testified, they could present hearsay. Summonses were also part of the legal arsenal but were not always successful. On at least one occasion, a summons could not be served, as the lobster fishermen had "left their homes for the woods."[72] The membership of the lobster commissions and inquiries were unlike the Rowell-Sirois Commission members described by historian Jessica Squires as seeing "themselves as solving a technical problem in a disinterested and decidedly nonideological manner." These men were deeply and personally involved.[73]

Those appointed to conduct inquiries were either employees of the Federal Department or identifiable partisans, almost always with some connection to the industry. A sitting member of the Federal or Provincial House or a failed or future candidacy were common qualifications. No fisherman appears to have been a commission or committee member, although William Whitman of the 1898 Royal Commission may have fished part time as a farmer.[74] Moses H. Nickerson, another provincial politician and member of the same Commission, never fished but had close relations with lobster fishermen as the founder of the Fishermen's Union of Nova Scotia. For most of his life, he was associated with the live lobster trade.

The commissions also diverged on who they spoke with. Early inquiries tended to survey fisheries officers (such as the Gordon inquiry) or so-called prominent men – always packers.[75] The importance of canning lobsters meant that until the end of the nineteenth century, what was almost always called the lobster industry was lobster packing. I have located no transcripts for the 1898 Royal Commission, but there were general comments on the quality of testimony from ordinary fishermen, and the Commission visited thirty-three communities in Nova Scotia, nine in New Brunswick, four in Quebec, and three in Prince Edward Island.[76] The number of communities reflected the disparate geographic organization of the lobster industry but also the political pressure and potential political benefit of consulting widely.

The transcripts for the 1903–4 inquiry into the lobster fishery in the Bay of Fundy and the Magdalen Islands document meetings with fishermen, as well as packers and priests. As mentioned earlier, William Wakeham's 1910 investigation purposely sought out the oldest fisherman in every community he visited and questioned them to determine change over time. The 1927 Royal Commission on the eastern fisheries also employed a consistent strategy to gather local knowledge. Every fisherman's testimony began with a statement of how many years he had been fishing; his experience established his credentials. Much of the testimony concerned lobster. The Maclean Commission heard testimony from men who had been on the water for nearly fifty years, capturing a moment just before the industrial groundfish fishery began to rise in importance. The Commission's origins were rooted in the same wave of populist political discontent that gave rise to the Antigonish movement – a Catholic-led social action campaign emphasizing adult education and cooperatives to improve economic conditions in local fishing and mining communities. Reflecting this context, the Commission made a deliberate effort to solicit the participation of fishermen. In every harbour, fishermen were instructed to hold public meetings, gather community recommendations, and select one or two local delegates whose travel and living expenses were covered if they testified before the Commission.[77] This strategy of consultation served dual purposes: to uncover local perspectives and to dissipate regional grassroots discontent by projecting the appearance of action. It also anticipated the participatory democracy associated with the 1960s and 1970s. In 1927 fishermen were expected to cultivate skills of democratic citizenship

through structured public meetings with designated chairs and sec-
retaries. Delegates were to be held accountable by having two men
(they were always men) report local recommendations, and financial
barriers to participation were eliminated.

The first major lobster commission was established in 1887 in response
to the sudden decline in the catch after a meteoric take-off. The mandate
explicitly asked for recommendations so that "the impending destruction of this Fishery may be averted."[78] Its focus was on government
regulation, and the position of at least some of its participants was well
known before the Commission began its work. J. Hunter-Duvar, Secretary to the Commission and Fishery Inspector for Prince Edward Island,
had already complained of closed factories and bankruptcies, the result
of an "exhausted" lobster fishery. As a journalist, poet, and novelist
Hunter-Duvar brought a rhetorical flourish to his reports, interventions,
and conservation arguments that his fellow appointees lacked.[79]

As the Commission was looking at a fishery in crisis, oysters were
tacked on at the last minute. The four commissioners were given a
timeline of six weeks and managed to spend almost twice the $2,000
budget the Commission had been originally appropriated.[80] The final
report generated a great deal of information about the material conditions and the structure of the canning sector but was ultimately split
between Commissioner Alfred Ogden, who believed lobsters could be
fished year-round, and the other three who supported regulations.

Half of the four-person commission was composed of defeated
federal Conservative candidates. Edward Hackett of Tignish, Prince
Edward Island, was appointed as chairman, holding the qualifications
of being a former Conservative member of the provincial legislature,
defeated federal Conservative parliamentarian, and his involvement
in the fish trade. After the Commission, he replaced fellow Commissioner Hunter-Duvar as Island Inspector of Fisheries, a position he held
until 1896 when he returned as Member of Parliament for West Prince.[81]
Another appointee, Ogden, as already mentioned, had been the Conservative Member of Parliament for Guysborough.[82]

While the 1887 commission questioned packers and suppliers, in
1889 Lieutenant A.R. Gordon of the Fisheries Protection Service surveyed another group of experts, government fishery officers. Gordon
gathered information on American ownership, the extent to which
local canners and fishermen adhered to regulations, price trends, and
what other kinds of employment lobster fishermen also pursued.[83] This
inquiry had a low political profile and was part of deliberation around
the introduction of a potential lobster fishing moratorium.

Federal politicians were also being prepared for a possible lobster
moratorium in the 1892 Special Report on the Lobster Industry, which

addressed the lack of general knowledge about this economically important fishery. This report, authored by the Fisheries Branch with no outside consultations beyond its counterparts in the United States and Newfoundland, established the economic importance of the fishery, repeated concerns about its "exhaustion," offered a history of the fishery in Canada and the United States, and defended conservation regulations with the position that "lobsters belong to the public not to the packer." The report addressed the impact of a potential moratorium on the fishery, arguing that women employed in small factories could find work in domestic service and farm labour and that the idle factories could be used for other products such as mackerel, fruits, and berries.[84]

The Liberals who profited from fishermen's electoral support in their rise to power in 1896 launched a Royal Commission on the Lobster Industry in September 1898.[85] The inquiry was framed, both in terms of the economic importance of the fishery and the claim that "existing regulations" were not "sufficiently stringent or elastic as to locality, for the proper conservation, protection and development of the lobster industry." E.E. Prince, Dominion Commissioner of Fisheries, chaired the commission of eight men, selected for Liberal connections with attention to geographic representation and one Acadian voice.[86] Local politicians lobbied for their particular nominees, and the Minister of Marine and Fisheries had a difficult time composing a commission that would not be weighted too strongly in favour of the packers.[87] He approached a Cape Breton Member of Parliament, requesting the name of "an intelligent fisherman or at least one in touch with fishermen that they would accept his report." This member recommended Donald Campbell, a farmer from Margaree Forks, who one disappointed packer claimed had never seen a lobster in his life.[88] Henry C. Le Vatte of Louisbourg was a merchant linked to packers, a local politician, and an "influential member of the Liberal Party" who subsequently received a patronage position as a Fishery Officer.[89] The other Nova Scotians were William Whitman, a farmer from Guysborough and future Liberal provincial politician who ten years later claimed he fished, and Moses H. Nickerson of Cape Sable, a journalist and future Nova Scotia Liberal MLA. From Prince Edward Island were merchant and later president of the Kings County Liberal Association Archibald Currie of Souris and merchant Liberal MLA Stephen (Étienne) E. Gallant of Egmont Bay.[90] Finally there was Patrick J. Sweeney of Shediac and Robert Lindsay of Gaspé. The partisan dividends were distributed beyond members who received $7 per day plus expenses, as the selection of hotel accommodation had to be politically approved. The result was a mix of evidence-based inquiry and partisan argumentation.

The Commission held sixty-five sittings throughout the region – but notably excluded Quebec's North Shore and the Magdalen Islands. These two areas were difficult to reach, especially in winter, and the omission of the Magdalens prompted a special inquiry there four years later. Of the communities visited, two-thirds were in Nova Scotia, while only three sessions were held on Prince Edward Island. There appears to have been some correlation between the locations selected and federal political representation: communities represented by Liberal members of Parliament were more likely to receive visits. This may explain the cluster of sessions south of Lunenburg in Nova Scotia and the absence of any meetings in eastern Prince Edward Island.[91] In Halifax, the testimony was weighted in favour of the canners, but in smaller communities, the fishermen were more likely to be heard. Still, despite the Commission's extensive regional travel, barriers for participation persisted. E.C. Bowers, a former Liberal MP and future member of the 1903 Commission, warned that the meetings would "be one sided" unless the commissioners went directly to where fishermen lived. As Bowers put it, "poor fishermen" could not afford a trip of even twenty or thirty miles to attend, and unless this was addressed, their voices would be excluded from the process.[92] The number of communities to be visited expanded somewhat, and Prince may have been surprised by the quality of evidence fishermen brought to the Commission. The Commission's report concluded that the fishermen's testimony had contained a great deal of "real information" obtained through observation, but they were also lacking "a large amount of information of which they should be in possession." The gap could be addressed with education.[93]

The Commission presented an overview of the industry at the turn of the century, noting that while nowhere was lobster the sole means of support, it was important almost everywhere. This was especially the case for the south and east coasts of Cape Breton, sections of Nova Scotia's Eastern Shore, Richibucto to Baie Verte on the Northumberland Strait, and in the Cape Sable Island area of Nova Scotia. The fishery was embedded in a context of occupational pluralism as lobster fishermen farmed, worked in the woods, or were engaged in other fisheries. It noted, however, that the highly remunerative character of the lobster fishery attracted many men who could not make the same money either farming or bank fishing.[94]

In the end, the eight commissioners failed to submit a unanimous report. Even the majority report had been contentious – particularly around the issue of size limits. Prince Edward Island merchant Currie opposed the proposed size limit, arguing that it would effectively end lobster canning in the Gulf. According to his Liberal colleagues, Currie successfully lobbied for a compromise: the recommended size limit would be reduced everywhere except west of Halifax.[95] Two commissioners, Whitman and Le Vatte, went further, refusing to endorse the majority report altogether. They argued that lobster stocks were not in serious decline and maintained that that it was impossible to overfish lobsters.[96] In contrast, the majority report held that there was a serious problem and government action was required. Not all the recommendations of the majority report were subsequently implemented, but conclusions of the 1898 Commission shape the fishery to this day. The final report noted that canning had reached its maximum growth and the future – already unfolding in southwest Nova Scotia – was the live market trade. It recommended that no new "foreign firms" be granted canning licences and that the idea of geographic monopolies granted to individual packers should not be pursued.[97] Artificial propagation of lobster eggs in government hatcheries was endorsed enthusiastically, as it limited no one's actions. Most importantly, the Commission recommended that changes be made to the two large geographic areas that defined lobster fishing seasons. Instead, the region's coast should be divided into six different fishing districts with distinct seasons and size limits that reflected local water temperatures, ice patterns, moulting, and reproduction cycles. In the original six districts, you can see the lines of the forty-seven districts that exist today.

Following the 1898 Royal Commission, a minor inquiry was established in 1904 to deal with a local Shippegan Peninsula practice among lobster fishermen. Here fishermen themselves cooked, shelled, and sold broken lobster meat directly to local canneries at a premium price. The Department of Marine and Fisheries appointed a local schoolteacher Pierre Morais of Caraquet, described as someone who had "practical knowledge of the grievances of the Fishermen and the conditions of the fishery," to investigate. He may have had too much practical knowledge, as he concluded that, as these poor men did not have the means to establish canneries for themselves, selling meat was a good strategy for increasing their earnings. This is an interesting example of someone on the ground, appointed to gather information, who made a recommendation at odds with the Department's policy of promoting increased sanitation in canning. The Fisheries Branch did not support this proposal, and the practice remained illegal. Here a local perspective was rejected.[98]

The absence of the Magdalen Islands from the 1898 Royal Commission meant that a 1903 department inquiry, which was originally intended to examine Bay of Fundy fishing issues, was expanded to include lobster both there and in the Magdalen Islands. The distance of the two locations and a winter season meant that weather again interrupted travel, and the inquiry took two years to report.[99] This Commission was chaired by a sitting Member of Parliament from Saint John, Colonel J.J. Tucker. Two other Liberal partisans, fish dealer and former Digby Member of Parliament Edward C. Bowers and defeated federal Liberal candidate journalist R.E. Armstrong of St. Andrews, weighed the five-person Commission in favour of the Bay of Fundy. The fourth and fifth members of this Commission were Robert N. Venning, Superintendent of Fisheries, and Reverend J.S. Turbide, a priest on the Magdalens. On the Magdalen Islands, commissioners met with fishermen, packers, and local priests, a pattern repeated, without the clergy on Digby Neck, the French Shore, and Grand Manan and Campobello Islands.

If the 1899 report had highlighted the natural intelligence of fishermen, the 1905 final report stereotyped the Magdalen Islands lobstermen as unable to recognize their own long-term interests.[100] In the end, the Commission recommended more rigid enforcement of the lobster regulations on the Magdalen Islands and submitted a divided report on recommended size limits for lobster in the Bay of Fundy. As the Commission dealt with other issues, only eight of its twenty-three recommendations concerned lobster.

The 1910 Commission on the Lobster Industry in Quebec and the Maritime Provinces sprang from the 1909 Select Committee of Parliament on Marine and Fisheries. This Ottawa committee focused on lobster, interviewing sixteen witnesses, and recording nearly 300 pages of testimony. Its April 1909 conclusion was that it required further information, and this was amplified a month later when Cape Breton lobster fishermen tied up their boats to protest low prices.[101] Fishermen were front and centre, and for the first time their words survive in some form, transcribed and often translated into English, when Commander William Wakeham, Officer in Charge of the Gulf Fisheries Division, began an extensive tour of the Atlantic coast. In the company of Superintendent of Fisheries Venning, and occasionally local Members of Parliament or members of the Select Committee on Marine and Fisheries, from June to September 1909 they visited fifty communities and recorded 1,240 pages of testimony from 167 fishermen and eighty-four packers. Wakeham had been in the field since 1879 and had served on various commissions. He brought to this commission extensive experience and direct knowledge.[102] What was remarkable about this inquiry was that the commissioners travelled by water and visited small coastal communities, not railway towns. Leger's Brook, Cape Bald (Cap-Pelé), Cape Tormentine, and Shemogue were called upon, not Shediac.

We know something of how witnesses were recruited as posters were sent to the postmaster of communities scheduled for visits and notices of meetings appeared in local newspapers.[103] Some individuals received personal invitations, and there were probably links between some local fishery overseers and the fishermen who testified. In Chester, Nova Scotia, the local overseer explained that the expected lobstermen from Tancook Island had not appeared as the herring had struck, and they were otherwise occupied. Fishermen were asked "where do you belong" and questioned about every aspect of the lobster fishery and how this changed over time.[104]

The testimony at the community level during the commission could hardly have contrasted more sharply with the evidence produced in 1912, when the new Conservative government established a Shellfish Commission. In this inquiry, lobsters were grouped with oysters and clams, as all three fisheries were in a "critical state."[105] The Shellfish Commission of 1912–13 focused primarily on the packing side of operations. Its members included two prominent lobster packers – Richard O'Leary of Richibucto, New Brunswick, and John McLean of Souris, Prince Edward Island – along with Halifax fish dealer S.Y. Wilson, who joined E.E. Prince.[106] This commission emphasized business concerns, and its witnesses were politicians or industry people. It had fifty public sittings, heard from 248 witnesses, and extended its investigation beyond the Maritimes to Boston and Rhode Island, key markets for Canadian shellfish. Its final recommendation on lobster size limits aimed less at protecting lobsters and more at preserving the packing industry.[107]

The 1927 Royal Commission to Investigate the Fisheries of the Maritime Provinces and the Magdalen Islands, under the chairmanship of Nova Scotian A.K. Maclean offers insights into the situation in the interwar period.[108] In the composition of the Commission, Maclean, president of the Exchequer Court of Canada and a former Liberal politician at both the provincial and federal levels, was joined by four Liberal partisans. Prince Edward Island-born Cyrus Macmillan was a McGill University English professor and Liberal politician. New Brunswick was represented by Jean George Robichaud, a fish dealer who served in both the provincial legislature and federal house. Finally, H.R.L. Bill, a Lockeport wholesale fish merchant and future Member of the Legislative Assembly, along with Joseph George Mombourquette of L'Ardoise, a former member of the Nova Scotia-appointed Legislative Council, represented that province. The origin of this Commission had much in common with the 1909 inquiry, as it was grounded in populist agitation and partially travelled by water on the CSS *Acadia*. While the broader context was the industrialization – and capitalization – of the offshore fishery, much of the Commission's testimony focused instead on the inshore lobster fishery and the early success of canning and marketing cooperatives.

Over the course of forty-nine hearings, the Commission heard from 823 witnesses and created 5,700 typewritten pages of testimony.[109] Those who testified included fishery guardians, overseers, packers, associational representatives, and members of the clergy, but the vast majority identified as ordinary fishermen. Notwithstanding the volume of words recorded, our knowledge of fishermen's concerns remains limited. Their voices were shaped and constrained by the format of the hearings, particularly by the structured questions of Counsel. Commissioner Macmillan, who in a letter to his wife identified himself as "the fishermen's friend," observed that despite the presence of witnesses, the fishermen "seem to be voiceless."[110]

Macmillan, the future inaugural Minister of Fisheries, who had just finished service on the Royal Commission on Maritime Claims, noted candidly to his wife his surprise that, for political reasons, the fisheries commission's delegation was larger than the one on which he had just served. His private personal correspondence offers insight into the Commission and alerts us to the invisible roles of Counsel and Assistant Counsel, the presence of senior civil servants who do not appear in the formal report, and the work of four male stenographers who travelled with the party.[111] We also see that there were opportunities to chat "alone" with fishermen after sessions and with prominent community men at informal lunches.[112] Macmillan's letters also contain Commission gossip. Portland Packing Company's powerful manager W.F. Tidmarsh did not testify before the Commission as, according to Macmillan, he "'beat' it out of town for he heard that some of the Inspectors had 'primed' the Counsel with questions to ask him."[113] In February 1928, after the Commission was finished with formal hearings, the president of the Nova Scotia Liberal Association travelled to Montreal from Ottawa to see Macmillan, as he was concerned that the report would be "concocted" by the Chair Judge Maclean and the Counsel. Rumours circulated that the report would "deal with small things like lobster seasons," issues that reflected testimony, and not "big things" like corporate concerns that were important to the Nova Scotia Liberal Party. In this personal conversation Macmillan acknowledged that he "had noticed some secret workings going on."[114]

The Commission's testimony gave unexpected weight to the decline of lobster stocks, the persistence of illegal fishing, and the harsh living conditions faced by inshore fishermen and their families. Its final recommendations reflected these concerns, calling for more science, greater bureaucratic oversight, and tighter regulation. The influence of fishermen's voices is most visible in the report's final paragraph, almost certainly penned by Cyrus Macmillan. In one of the most poetic paragraphs ever to appear in a Royal Commission report, it evoked the lives of "of fisher-folk despondent and disheartened, struggling on against economic disabilities, eager to labour in one of the most hazardous of pursuits but unable to sell their products for a reasonable reward, always hoping for better luck, and clinging grimly and patiently to their calling – a tribute at once to their character and their courage." It continued to call attention to the tragedy that the next generation appeared to be making different choices and were already "planning to migrate at maturity to another land to make a living."[115] Emotive and human, the Commission's conclusions represented a very different kind of knowledge from that of businessmen, scientists, and bureaucrats. Rather than electoral calculation or industrial efficiency, its focus turned towards social renewal. It ended not with technocratic solutions, but with a call to reimagine the fishery through cooperative organizations.

The Commission reflected one of the last moments until almost the end of the twentieth century when recognizing community-based knowledge was politically useful. A period of diverse, although admittedly uneven, input from various stakeholders was coming to an end as authority concentrated around specialized research scientists and, later, the theoretical models of economists. While some pre-First World War commissions were dismissive of the expertise of fishermen, there are other examples of traditional ecological knowledge being solicited and respected. In moments of organized agitation or in partisan efforts to capitalize on fishermen's electoral support, there were seeds of more inclusive democratic consultation. This is important, as the lobster industry was a labour-intensive rather than capital-intensive industry and these consultative processes moderated the power and influence of canners and packers. As discussed in chapter 6, the geographic and market fragmentation of the fishery meant that it was rarely possible for fishermen to organize across significant divisions, and lobstermen diverged in their attitudes towards regulation. These commissions, however, created opportunities for decision-makers to interact with fishermen, made room for experience-based knowledge, and on the best days extended empathy and respect.[116]

Community-based commissions contrasted with the expansion of laboratories and full-time fishery biologists under the Biological Board of Canada, renamed the Fisheries Research Board in 1937.[117] As the location where knowledge was created changed, so did strategies about how information should be deployed. In the interwar period, greater emphasis was placed on saving the lobster stock through education, rather than regulations alone. In 1912 the former schoolteacher and now Superintendent of Fisheries W.A. Found had made one of the earliest public arguments against hatcheries, stating that if they closed, an educational campaign for fishermen, emphasizing "their own personal interest" would better protect berried lobsters.[118] When hatcheries closed in 1918, the Fisheries Branch launched an educational program on lobster biology with an eye to conservation. Under the direction of A.P. Knight, with an unidentified francophone counterpart from Université Laval responsible for eastern and northern New Brunswick, lecturers travelled throughout the region giving formal presentations in local halls, schools, and churches and less formal talks in canneries and on wharves.[119] In 1920 Andrew H. Halkett, the department's "naturalist," spoke in thirty locations between Chester and Digby Neck to an estimated 1,200 Nova Scotian lobster fishermen in a formal talk accompanied by lantern slides and "open discussion."[120] This was a turning point, as experts were not gathering information from fishermen or packers; rather, scientists and bureaucrats claimed exclusive expertise. This new strategy also placed primary responsibility for the conservation of lobster on individual fishermen, the most vulnerable participants in the industry, not the packers or buyers.

In 1940 a scientific expert with formal academic credentials, and with the unlikely name of Donovan Bartley Finn was named Deputy Minister of the Department of Fisheries.[121] Donovan Finn, PhD, led a department that became even more removed from the experience of those who fished. Even experts who worked with living beings in laboratories or in controlled scientific experiments were sidelined by the abstract modelling of economists.

Conclusion

In the late nineteenth and early twentieth centuries, Ottawa's "lobster question" posed an impossible challenge: What policy could both protect and exploit lobster, secure its long-term sustainability, and deliver immediate economic returns with short-term political dividends? Any viable approach also had to navigate the pressures of local electoral politics and the structural tensions of federalism. Unsurprisingly, the Department leaned heavily on the rhetoric of impartial science to justify politically popular but ultimately unsuccessful initiatives – such as government lobster hatcheries or the ill-fated attempts to introduce lobsters into the Pacific Ocean. Even in its weakness, state regulation reflected what land-locked Ottawa *knew* – or believed it *knew* – about the lobster fishery. The economic importance of lobsters to numerous distant Atlantic communities led those in charge at the Fisheries Branch or Department to place faith in information as a source of legitimacy. Better data, it was hoped, would lead to better policy, improved enforcement and perhaps greater acceptance of the state's role. Before the Second World War, the knowledge underpinning policy was largely practical and empirical, and there was space, albeit limited, for experience-based local knowledge. This is most evident in the counting: cans, boats, traps, canneries, and fishermen. It was also evident in the use of commissions and inquires to gather information from Fisheries employees, packers, and eventually fishermen themselves. The agendas being pushed from Ottawa – whether aimed at extracting votes, quietening regional discontent (both Atlantic and Pacific), or mediating the conflicting interests of fishermen and packers – were rarely neutral. Yet information gave the state the illusion of objectivity. As we shall see in chapter 5, this illusion buttressed the little authority and legitimacy the federal government could claim in managing a resource it only partially understood.

Chapter Five

Rejecting Regulation: Direct Action, Force, and New Strategies, 1890s to 1960s

Although formal laws and regulations governing the lobster fishery had existed since 1873, there is a long history of canners and fishermen openly defying them and rejecting their legitimacy. A 1900 editorial in the *Yarmouth Times* observed that there were not twenty-five fishermen in the rich lobster district between Chebougue Point and Barrington "that are observing the law in any of its forms today. Where in the past, I don't think there were that many poachers."[1] As noted earlier, both fishermen and packers had a strong sense of the "right" to fish, grounded in longstanding custom and economic necessity. Lobster regulations, though legally binding, often lacked social legitimacy or "buy in." Local political partisan activity could extenuate this dissent. Community-level patronage appointments were an important aspect of how Canada was governed, but this did not reward competence or rule-following. For most of the period before 1970, the Canadian state lacked the coercive capacity to enforce its own fishery regulations, both political will and surveillance technology were insufficient, rendering many formal regulations aspirational at best.

Lobster fishing and production were extremely decentralized, and there was neither the political consensus nor the physical capacity to control the nearly 11,000 kilometres of coastline in the Maritimes alone. There was also a great deal of money and employment at stake. In 1961, for example, there were 21,000 lobster fishermen working in an industry worth $16 million. That year, one employee in the Fisheries Protective Service observed that there was "No other single industry" than lobster where so much money could be made locally in such a short time.[2] With money, time, and the number of fishermen and packers involved, it is hardly surprising that illicit activity persisted, albeit unevenly in terms of time and place. As noted earlier, informal customs and practices mirrored the formal law. Here we see that they were also paralleled by informal forms of justice as individual fishermen or local communities capitalized on the weak coercive power of the state to achieve their own ends. This chapter explores how and why enforcement was limited and how haphazardly enforced laws interacted with informal forms of regulation and justice. It covers a broad swath of time with the intention of emphasizing continuity rather than change. Of course, there was change between the 1890s and the 1960s. But the decentralized structure of the inshore lobster fishery, combined with the limits on surveillance and enforcement capacity of the state, constrained what formal laws could achieve. Just as importantly, deeply held local understandings of entitlements often trumped the authority of federal regulation, creating patterns – even cultures – of resistance. At the core of this resistance was economic necessity: illegally caught lobster remained a valuable income. Yet resistance was not absolute. Because the commons had a local context, fishermen and packers frequently turned to the state to enforce laws that aligned with their specific interests at particular moments, even as they evaded other fisheries regulations.

Generally, the emphasis on enforcement changed from focusing on packers in the nineteenth century to policing fishermen in the twentieth century. This shift in emphasis coincided with the reorganization of the industry from canning to live sales. An increased focus on individual fishermen framed illegal activity as both a workplace crime and a folk protest, something akin to rumrunning or other forms of rural wildlife poaching.[3] The lobster industry unquestionably had a longstanding reputation of operating outside the law. A 1921 *Maclean's* magazine article referred to a lobster fisherman as "the lobster pirate" and concluded that policing was impossible "except with the thorough co-operation of a majority of the fishermen themselves."[4] Similarly Richard Williams, writing on behalf of packers at this time, acknowledged the industry's dubious reputation. With rhetorical flourish he concluded that a Charles Dickens could write some stories that would startle readers, and an Upton Sinclair might write another *Jungle* about the moral and sanitary conditions existing in early factories.[5]

Certainly, some fishermen disregarded regulations aimed at conservation, cheated canners to whom they sold their goods, and occasionally – very much at their peril – stole from a competitor's trap or broke informal understandings of territorial rights. Eggs were washed off lobsters and undersized catches were stretched to size. If lobsters were sold by weight, pebbles were placed into joints, or if they were sold by count, large and medium lobsters were mutilated to make numbers appear greater.[6] Violence was also directed against fisheries officers and neighbours suspected of informing. It is held that at least one fisheries officer was murdered while pursuing poachers and one department official in the 1960s recounted illegal lobster activity as being linked with "innumerable hand-to-hand fights, shootings, knifing, backwoods chases, ambushes and home burnings." According to the same official, those tasked with enforcing the law, or neighbours who informed on fishermen could be subjected to arson, vandalism or be "socially ostracized, their children bullied."[7]

Canners also broke the law when they packed out of season, bought undersized or berried lobsters, accepted broken meat already removed from the shell, and sold canning supplies to bootleg operators who would sell them their products in return. Illegal canners operated everywhere throughout the Maritimes, as it was one of the only significant ways to profit from out of season fishing.[8] Although it took place everywhere, certain areas were infamous, such as the Eastern Shore of Nova Scotia, Cape Tormentine in New Brunswick, and the Egmont Bay district of Prince Edward Island. These activities not only broke the law, they also jeopardized the quality and reputation of canned lobster. Fisheries officials noted that Cape Tormentine was "notorious for packing a bad quality of lobster due to poaching after the close season."[9] Illegal operations operated blatantly in public or more discretely in shanties, woods, beaches, or houses. In the case of homes, elaborate decoys could be established with boiler pipes tied into chimneys to hide what was happening inside. Bootleg operations were almost impossible to catch in action (sometimes cans were seized that were still warm!), as elaborate flag warning systems and later telephones offered notice of where fishery officials might be going.[10]

Policing Regulations

The disproportionate efforts of the Fisheries Protective Service (FPS) on the Atlantic coast and specifically in connection with the lobster fishery has already been noted as a source of frustration for British Columbians. The FPS comprised water-based patrols originally intended to protect Canadian territorial jurisdiction against foreign encroachment. It owned a varying number of its own vessels but also annually chartered boats from political friends of the party in power. By the 1890s its reports noted that it was spending increasing amounts of time not policing American fishery schooners but enforcing federal lobster regulations.[11] This focus in the early 1890s reflected both the growth of the lobster fishery and a policy change between 1888 and 1894 when Charles Hibbert Tupper, as Minister of Fisheries, placed an unusually high priority on enforcement. In 1891, Tupper acknowledged the limits of state enforcement and introduced new legislation that focused on regulating packers through licensing as of 1894. In justifying this policy direction, he wrote that under the current lobster regulations there were so many "opportunities for evading the law" that enforcement would require "an army of officers distributed over the country."[12] But even in periods of supposedly rigorous enforcement, there was not even a clear understanding of the law by those charged with enforcing it. In 1894, for example, the Deputy Ministers of both the Departments of Justice and Marine and Fisheries had to weigh in on the matter of whether the fine for catching an undersized lobster was to be applied per lobster or for the entire offence, as opinions differed.[13]

In addition to the FPS, the Fisheries Branch operated an "outside" service, the category of people not based in Ottawa. Each province was divided by districts with permanent inspectors, who were part of the civil service. They supervised overseers, who held annual appointments. Both inspectors and overseers were "Fisheries Officers" and held magisterial powers on matters related to fishing regulations. This meant, like other relatively minor officials of the state, such as Indian agents, that they were ex-officio justices of the peace and in the extraordinary position of being able to prosecute, judge, and punish. In particular, they held sweeping powers over the destruction and confiscation of private property. Inspectors and overseers were given authority to "search, break open and search, or grant a warrant to search any house, vessel, or place where he had reason to believe that any fish taken in violation" of the *Fisheries Act* or its regulations might be located.[14]

In the early 1890s, a third class of officials – the permanent fishery wardens – were generally dismissed and replaced by special guardians who were paid as needed on a per diem basis. Guardians were most obviously political patronage positions and hired upon the recommendation of local patronage committees. The 1894 Annual Report of the Fisheries Branch reported 200 guardians employed as part of the "outside staff." By 1904 there were 434 fishery guardians employed across Canada, and the reported number peaked in the 1908–9 Department annual report that stated 680 guardians were engaged by the Fisheries Branch.[15] In 1912 guardians received between $24 and $50 a year, and the Prince Edward Island manager for the Portland Packing Company made no distinction between Conservative and Liberal practices when he described guardians as "generally small local politicians ... [who] look upon the amounts paid them as a reward for past services rather than a payment for work to be performed."[16] During the mid-1930s the widespread negligence of fisheries officers in Prince Edward Island became the subject of an official inquiry, and Commissioner Arthur T.

LeBlanc concluded that it was common for guardians to gain appointment through the influence of local lobster poachers.[17]

Overseers, for whom fisheries work was usually only a part-time occupation, were also generally political appointments, and a New Brunswick district inspector subtly acknowledged this in the 1898 Fisheries Branch annual report. The year after Conservative appointments would have expired following the 1896 Liberal federal victory, he explained that he had not presented any local reports from overseers as they had "no value," "as many of the local officers have been recently appointed."[18]

The legal power of the officers was limited by technology. Access to better vessels was an issue before the First World War. It has already been noted that a Southwest Nova Scotian overseer petitioned his superior in the Fisheries Branch in 1911 to provide small motorboats to guardians. The local lobster fishermen were all using power boats, but the officers were still rowing dories. Even though the cruiser *Arleux* reported destroying 1,700 illegally set lobster traps in 1936, this steam-powered former Royal Canadian Navy vessel was hampered by its maximum speed of around twenty kilometres an hour.[19]

The general pandemonium around enforcing regulations is evident in moments such as July 1898, when the Fisheries Branch sent public notices to local newspapers to inform fishermen that the lobster regulations would be enforced – and that guardians who failed to do their jobs would be dismissed.[20] The fact that such a notice was necessary reveals just how far the practices on the water diverged from official rules. This was not a new problem. Four years earlier, in August 1894, the Prince Edward Island Inspector had been explicitly instructed to report local officers who were not doing their job. Nor was it one that could be easily resolved. In January 1903, the Fisheries Branch took the extraordinary action of notifying officers that the law was to be enforced everywhere without favouritism.[21] The need to make this declaration certainly suggests that enforcing the law and regulations was not the status quo.

Yet it would also be a mistake to overlook the real authority these minor state officials held. Officers in charge of lobster patrol boats, for example, exercised prosecution and judicial powers. They regularly confiscated and destroyed property such as traps and ropes with no accountability. Seized boats were not destroyed but posed a different problem. They could not be sold locally out of fear that the boat would be stolen back by its owner. Here the lawbreaker clearly rejected the state's authority to take his property. In 1898, one Prince Edward Island fisheries officer suggested trying to sell a seized Beach Point boat in

Charlottetown, as "no one would know where she came from."[22] Beyond the destruction of traps, A.C. Bertram, Inspector of Fisheries responsible for Nova Scotia's Eastern Shore, recounted his "wholesome lesson" delivered to a group packing out of season that included both arrests and burning down two buildings housing canning equipment at Whitehead, Guysborough County.[23] This attack on private property was done without any need for due process.

The organization of the outside staff changed over time in a manner consistent with general civil service reform. In 1918 local policing was restructured into a smaller group of better-paid local overseers. This "intelligent group of young active men" was given formal training and incorporated into the civil service to be supposedly free of political interference. The 1918 annual report acknowledged that the "mere pittances" earlier overseers had been paid meant that it had been unreasonable to expect them to work full time.[24] By the 1921–2 annual report, there were 300 fewer guardians, but these positions remained political appointments. Indeed, it is difficult not to snicker at the grandiose claim made a year later in the department's annual report that, with the new system instituted in Prince Edward Island, "there is no doubt that illegal lobster fishing in this Province will be a thing of the past."[25]

James G. d'Entremont of West Pubnico, Nova Scotia, began as a fishery overseer in 1912 and held the position for twenty-five years. He recounted on his retirement that the job began with a "small fixed annual salary" and mileage for a horse. In 1914 he was the first overseer in the Maritimes to use an automobile, and in 1929, after a six-week course, d'Entremont was promoted to inspector.[26] Throughout his long career, he was deeply embedded in the local community (d'Entremont is the predominant surname in the Pubnicos). His family and neighbours were the fishermen he was supposed to police and prosecute. The fact that those responsible for policing were so deeply connected to their communities was recognized as a problem, as one inspector complained in 1912 of "the unwillingness of Fishery Officers to bring action against their neighbours."[27] D'Entremont, at least had some employment security, but guardians who were hired by the day had even more to lose with alienating family and neighbours. Brydon Smith who was employed as a seasonal guardian in the early 1930s testified that he "did not want to catch my friends who I had poached with. There was little use making enemies over two months."[28]

The enforcement of lobster regulations was impossible work, if taken at all seriously. As early as 1893, FPS officers complained that finding illegally set traps was "very slow business" because they had to resort to dragging the ocean's bottom, as fishermen lobstering out of season did not use buoys but located their traps by landmarks.[29] Moreover, although traps and rope might be confiscated, fishermen themselves escaped punishment, as they rarely were caught in the act of fishing their traps. But even the confiscation and destruction of traps could lead to retribution. In April 1895 the steam tug *Wenela*, used for patrolling illegal lobster fishing in the Northumberland Strait, was set on fire in an act assumed to be "vengeance on the Department" for its work.[30]

Even on land, the task of enforcement was next to impossible. Obviously, it was impractical to check every lobster that landed at a cannery for size or the presence of eggs, and spot checks were hampered by signals used to relay information about the whereabouts of officers. Fishery guardians were well known and easily identified when travelling by land. Notice of their approach was broadcasted from factory to factory, and much bootleg canning took place off roads in bushes or swamps.[31] As a result, when raids were successful, they were often under the cover of darkness. In 1886 a midnight raid of lobster canneries at Sandy Point and Gunning Cove in Shelburne County resulted in the seizure of undersized lobsters.[32] Author Evelyn Richardson recounted stories she had heard of her grandfather in the same area and probably about the same time, evading the local inspector when searching for undersized lobsters called tinkers. She described a system where all factories had warning systems (her grandfather used a blanket hung in a cookhouse upstairs window) to warn fishermen not to land if an inspector was about.[33] In the 1950s in New Brunswick, red and green lights were used to signal poachers of the presence of fisheries officers, and in Cap-Pelé a flag was still used. There was also a long tradition in some communities of hanging laundry in a certain way.[34]

After the Second World War, the Department of Fisheries hired veterans with war service experience to serve as fisheries officers. Men such as Isaac Vigneault, who held this position from 1947 to 1978, worked between Pictou Island and Pugwash on the Northumberland Strait. Vigneault was a veteran who served in Europe and Africa and, as part of the "class of '47," participated in the one and only specialized course offered at the Technical College of Nova Scotia for new fishery officers. In the early years Vigneault was responsible for licensing, collecting statistics, and enforcement, along with inspecting canneries and fish plants in his jurisdiction.[35]

By the early 1950s around 200 uniformed protection officers and part-time wardens, eight members of the mobile squad, and twenty-two coastal patrol boats with an occasional chartered aircraft were charged with enforcing lobster regulations. There was an increased emphasis on prosecution despite the "apathetic" support of magistrates and the general public, alongside the introduction of new educational campaigns.[36] Indeed, there is plenty of evidence that local magistrates did not take lobster violations seriously. In June 1954 a Prince Edward Island fisherman pleaded guilty to possessing 420 "shorts" and was only fined $5 for breaking the law. Even intensified enforcement campaigns were uneven. The department reported 440 lobster prosecutions in 1953, with 90 per cent in western Prince Edward Island or Southeast New Brunswick.[37]

Illegal lobster fishing was the most regular theme connected to the fishery in the industry trade journal *Canadian Fisherman* from 1954 to 1959. Its critical position was at odds with the court of public opinion that often sided with those breaking the law. The legitimacy of regulations was still not universally accepted. In 1962 a report from the St. Francis Xavier University Extension Department noted that "public opinion is largely in favour of the poachers," as local customers, "even people in very influential position," enjoyed buying illegal lobsters at lower prices. One poacher was quoted as saying there was no such thing as illegal lobsters, as the government "didn't put them there and no fisheries officer is going to tell us when to take them out."[38] Much earlier in 1903, A.M. Hatfield, an overseer based in Yarmouth County, described a world view that was both ambivalent and pragmatic, as local fishermen usually accepted fines "without complaint," but if they were not caught they had "more bread for the children." According to Hatfield, fishermen did not see catching undersized lobsters as a "heinous crime," as one could catch a lobster in December that was a quarter-inch short and catch the same lobster in March, April, or May and it might have reached the legal limit.[39]

In 1954 the Department of Fisheries established a specialized mobile squad of enforcement officers – sometimes referred to as the "Goon Squad" – who could be sent to specific lobster trouble spots, as members had no personal pre-existing social connections to the local community.[40] This modern militarized approach was used along with broader department reforms. The use of radio communication on patrol boats, faster cutters (but not fast enough to outrun new high-powered outboard motors), the occasional use of chartered aircraft, and expanded judicial authority seemed to, at least temporarily, increase efficiency.[41] A veteran Royal Canadian Airforce pilot in 1948 was probably the first use of aerial surveillance when he borrowed a plane to spot submerged crates near Alberton, Prince Edward Island, from the air. Poachers responded by painting their traps and crates seaweed green as camouflage.[42]

New Glasgow-born Stanley Dudka, another war veteran, joined "the Goon Squad" in 1959, and on his retirement in 1988 he recounted that he had been "shot at 18 times, punched, beaten, lured into fights and suffered numerous bruises plus a broken rib." His boat was rammed and split in two at sea, he was blackmailed, and he survived an organized campaign to remove him from his job.[43] As historian Bill Parenteau has demonstrated elsewhere, fishery inspectors and guardians had little support or sympathy among the public. In 1898 Inspector R.A. Chapman reported to his superiors that Prince Edward Island "fishermen are terribly bitter at me, especially for searching houses." These searches yielded results such as the seizure that same year of 107 cans of lobster hidden in a bed in a Nova Scotia Eastern Shore home.[44]

Indeed, the work could be dangerous and fishery officials were occasionally shot at or assaulted – a problem particularly serious in southeastern New Brunswick and the Miramichi. In October 1926 fishery officer Agapit LeBlanc was murdered at Buctouche while probably investigating lobster poachers. He continued despite being warned off, and the case was never solved.[45] In the 1930s a former fisheries official based in Bay Du Vin with twenty years of experience was shot at while trying to liberate lobsters from illegal traps at Pointe-Sapin. He abandoned the efforts that night, fearing the eight men gathered around his car might push it over the bank.[46] The guardian later testified that he was afraid to go to Pointe-Sapin at night, as he might be "shot or stoned to death." After the war in 1950, violence escalated when intruders burned the barn, crops, and livestock of Antonio Turbide, a fishery officer at Eel River Bridge, New Brunswick, while he was at work.[47] More incidents followed in September 1958 in southeastern New Brunswick, including reports of fishery officers being shot at in their patrol boats and cars and one knife attack.[48] In 1960, in an unidentified village, fishery wardens were met with stone-throwing "even" by women and children." The following year two wardens were assaulted by a group of men; and one suffered serious head injuries and was hospitalized for over a month.[49] While southeastern New Brunswick had a particular reputation for violence, it was not unique. In 1904 a shot was fired at the local fishery officer in West Green Harbour, Shelburne County, as he investigated illegal lobster fishing.[50] A year later, at Port L'Hebert, a fishery warden discovered traps set just before the season started, and while cutting a few of them adrift he was assaulted by a fisherman – who according to the local report "got the worse of it." There was much community sympathy for the fisherman.[51] Far from being isolated incidents, these episodes reflected a broader pattern of confrontation, where fishery officers were not only undermined but sometimes openly targeted for fulfilling their duties.

The same kind of violence or property destruction was directed at community members suspected of informing on illegal activity. An informer at White Sands, Prince Edward Island, in 1898 found his boat had seventeen holes bored in the bottom. As the local overseer explained, "people are rather spiteful down here."[52] Neighbours informed on neighbours, and anonymous notes have found their way into various archives. One such note from October 1927 claimed that a named Riverport, Nova Scotia, fisherman, his sons, and – rather exceptionally – his wife had thirty traps set out. They were accused of selling cooked lobsters in town out of season and "making as much as any man in the spring." The note concluded, "I wont put my name on this but you can be shore it is true."[53] An informer wrote to the Newfoundland Minister of Fisheries and Co-operatives in 1952 to let him know that his neighbours were setting traps early and emphasized that he dreaded "being branded an informer" and that "if it should be suspected then I am finished around here."[54] In 1954 two fishermen from Pointe-Sapin were convicted in another case of arson after burning down the house of the man they thought had informed on them. At nearby St.-Édouard-de-Kent, New Brunswick, in the autumn of 1961, twelve poachers invaded the home of a fisherman they accused of snitching and threatened to burn his house and his boat.[55]

Canners

While the previous section concentrated on fishermen, it needs to be emphasized that canners were also remarkably effective at evading the law. Large, legitimate packers used and depended on small bootleg operations for products during periods when they could not meet market demands. At other times, when oversupplied markets increased competition, large legal canners led efforts to squeeze out smaller concerns through the expansion of state regulation using strategies of increased surveillance and heightened technological requirements.

Few illegal canneries could exist without supplies such as cans and the official stamps provided by legitimate packers. Nor could their products enter the market chain without legitimate wholesalers' assistance. This makes it almost impossible to make any differentiation between "legal" and "illegal" canners. Dealers found with illegal cans almost always refused to name their origins, suggesting that loyalty across legal and illegal operations was a necessary part of business. Indeed, there was little reason to turn over information as prosecutions were rare, and subject to political interference.[56] In 1921 Inspector Sylvain Gallant of Prince Edward Island relayed that illegal canning in the past "had been encouraged by politicians and dealers. Even clergymen had intervened on behalf of lawbreakers who pleaded poverty as an excuse. Election years were the worst for illegal packing."[57] Those in the employ of the Department of Marine and Fisheries could also be involved in illegal operations. Hector McKenzie of Flat River, Prince Edward Island, was found openly operating an out of season cannery with his brother who was a fishery officer. Another Tignish fishermen admitted to catching and canning about 65,000 pounds of lobster in the close fall season while holding a patronage position. As someone with a plum government job as a lighthouse keeper, he acknowledged that this had been an error in judgment.[58]

The amount of goods illegally packed varied from year to year as large companies used illegal operations to meet or stabilize market demands. In 1897 fisheries officials concluded that more was being packed in Guysborough during the close season than during the legal one.[59] Discoveries such as this led the Department of Marine and Fisheries in the 1890s to understand illegal packing as more harmful to the fishery than illegal fishing. Without illicit packing there would be no market for lobsters. Regularly large firms turned a blind eye when their local managers furnished small operators with supplies. When caught, these firms claimed that fishermen stole cans from their closed factories – and it is entirely possible that this might have occurred.[60] However, canners were rarely innocent. Writing about problems in Baie Verte, New Brunswick, during the 1905 season even after thousands of traps and miles of rope had been destroyed, the Dominion Commissioner of Fisheries noted that illegal fishing continued, as it was "countenanced and encouraged by the packers and large merchants." In the first decade of the twentieth century, fisheries officials in Prince Edward Island acknowledged that there was no attempt to enforce the minimum size limit there – nor had there been for some time; if regulations were enforced, the canneries could not operate.[61]

Legal operations were tied to illegal ones with extended credit. Medium-sized operator H.W. Longworth of Prince Edward Island approached Conservative Senator Donald Ferguson in October 1895 to plead the case of John Canteles, who operated a small lobster factory at Flat River. Longworth supplied Canteles with his cans, fishing supplies, and paid his $10 annual canning licence fee. When he was caught packing illegally, Canteles was fined and his pack was seized. In turn, Longworth paid the $25 fine, and asked that the illegal pack be returned to him as the one who had born all the expenses.[62]

It was regularly argued that fishermen broke the law in order not to starve, as if canners and suppliers had nothing to do with it. The

Charlottetown Guardian reported the story in May 1937 of Bessie Kinch, a mother of fifteen who had to "poach or starve."[63] Similarly, Alberton fisherman Arthur Wilkie testified before the 1937 Inquiry that cod fishing "did not make enough to pay for the salt for my porridge," and without lobster poaching his family may not have starved but "they would see plenty of meal times come around with no meals."[64] The same 1937 inquiry was informed that hard times and low prices meant "every fisherman" in the Tignish area had fished lobsters out of season in fall 1936 and that the local Imperial Oil agent had accepted illegally caught lobsters as payment on accounts.[65] Illegally packed cases of lobsters were sold to the Portland Packing Company and the local fishermen's cooperative. In addition, both firms moved live lobsters by truck at night from the north side of the Island to the Northumberland Strait where the season was still open.[66] The Tignish manager of the J.H. Myrick cannery testified that he mixed cases of legal and illegal lobsters in his shipments to wholesaler Roberts, Simpson and Company in Shediac and that he regularly reused transport permits. Canning also took place by independent agents in camps in the woods or in private homes. There is little need to speculate how George "Poach" Adams of Alberton South received his nickname as he canned in his home the lobsters he fished. To avoid detection, he had the pipe from a boiler used at the rear of the house routed through his kitchen and connected so that the steam would escape through the regular chimney.[67] In the 1930s poaching lobster was embedded into the entire Prince Edward Island supply chain. Fishermen in western Prince Edward Island were supplied with discounted bait and gasoline in exchange for out of season lobsters. The 1937 Commission recorded that local poachers were sold gasoline from New Brunswick at 16½¢ a gallon, while the standard local price excluding a fishermen's rebate was 30¢ a gallon.[68] The access to discounted gasoline tied fishermen to illegal buyers beyond the sale of their catch.

Into the 1960s illegal packing of undersized lobsters could be highly profitable for canners. Unofficial Department of Fisheries estimates suspected that between 10 and 20 per cent more lobsters were either caught out of season or undersized. In order to remain in business, legal canneries bought from poachers. For example, one large operator was convicted almost yearly between the Second World War and 1960. Department of Fisheries officers observed that these canneries simultaneously lobbied for increased enforcement and at the same time encouraged their buyers to acquire whatever lobsters they could get regardless if they were undersized or out of season.[69] In 1961 three pounds of live lobsters made one tin of twelve-ounce lobster meat drained. Fishermen were paid 32¢ a pound for legal sized canners in the shell, 18¢ per pound for undersized and illegal lobsters in the shell, and $1.20 a pound for shucked meat from undersized extra-legal lobsters delivered directly to the processing plants. All tins sold for $1.50, yielding a profit of 7½¢ per tin from legal lobsters, 49½¢ per tin from undersized canners, and 28½¢ per tin from illegal shucked meat. Not only was the profit margin significantly higher on the illegal catch but the volume provided by the undersized lobsters made packing viable for small operators.[70] In a May 1959 raid, thirty-two houses in Alberton, Prince Edward Island, were stormed in an early morning swoop. All houses had equipment to process and can lobsters, including can covers with the embossed number of two different legal licensed canneries. Around 800 pounds of canned lobster meat was discovered. Newspapers reported Mrs. Earle Callaghan had a "complete cannery" in operation in her kitchen, including more than 500 undersized lobsters. The raid was not without controversy, as children were scared "nearly hysterical" and officers at the door were described by locals as like "wartime Nazis."[71] It would be almost impossible to identify a more powerful image of the coercive state in post-war Canada than this label.

Sometimes illegal actions may have been in a grey zone. As already noted, around 1900, on the North Shore of New Brunswick, fishermen supplied canners not with live lobsters but with cooked, broken meat. Here the lobstermen took their catch home, boiled the lobsters there, and removed the meat.[72] So-called pound fishermen received a higher price than those who delivered their live catch to the packers' wharf, and undersized lobsters could not be identified. This system had the added benefit of allowing these part-time farmers to retain the shells for manure and advantaged packers who did not have to pay labourers for periods when there were no lobsters to process at their factories. Men such as Gervais Therriault of Caraquet explained that fishermen in his community were "not rich enough to put up regular canneries" and so conducted this illegal practice.[73] These men did not consider themselves to be breaking the law; rather, they were taking advantage of entrepreneurial opportunities available to them. The practice of selling "shucked meat" directly to canners re-emerged in the late 1950s and 1960s, as there was no way to link this meat to undersized lobster.

The limits of state coercion or state capacity to enforce the law and regulations was made explicit in various attempts to propose rules that could be enforced. In his 1910 report, long-time Fisheries Inspector William Wakeham attempted to reduce the ten lobster fishing districts seasons to seven, arguing that his recommendations were "practical and plain ... possible to enforce, with minimum of friction and cost."[74]

Three years later Prince, who had been a strong proponent of tailoring the fishing season to local conditions, reversed his position and now called – unsuccessfully – for a "universal and simultaneous close season along the Atlantic Coast." In this pragmatic proposal to eliminate district boundaries, Prince stated that a return to the pre-1899 two-part division for the entire Atlantic region was ideal but conceded that even five districts would be an improvement over the ten different districts to police. His goal was to eliminate as many boundaries as possible and was consistent with the advent a more "universal," efficient, liberal regulatory state. Attitudes towards enforcing size limits also changed over time. A minimum size was in place between 1874 and 1910, but it was largely unenforced. W.A. Found (future Deputy Minister of Fisheries) recommended in 1910 abandoning size limits (with the exception of Charlotte County, adjacent to Maine) and rather focus on the regulations requiring that female lobsters carrying eggs be set free where there was already some compliance by fishermen and packers.[75] With the exception of the Bay of Fundy area, size limits were generally abandoned, only to be reintroduced for the Atlantic coast in the 1930s where many fishermen were already participating in the more remunerative live market trade and interested in selling larger lobsters. After 1940 size limits were again extended to the Gulf of St. Lawrence. A law that could not be enforced undermined the department's authority and exposed its weakness.

The issue of open and close seasons and related boundaries also exposed the limitations of the state's power to enforce laws and regulations and offers glimpses of the constraints on its authority. Two cases on each side of District 8 on the Northumberland Strait – both in the 1930s – illustrate this complexity. Few places attracted more attention and caused more frustration for fisheries officers than "the Chockpish Line," sometimes reported as "Chock Fish" in government documents. This dividing mark referred to New Brunswick's Chockpish River located across the Northumberland Strait from the western end of Prince Edward Island. After 1899, it became the official boundary between fishing districts 7 and 8. Here, the Strait is only twenty kilometres wide, and on both shores lived relatively poor Acadian households, who combined lobster fishing with farming. While different districts possessed different standards of community compliance with regulations and undoubtedly different levels of policing, on the New Brunswick side fishermen around the Chockpish line were labelled by fisheries officials as outlaws and poachers. The 1899 lobster regulations gave the area north of the river a fishing season from 25 May to 10 August and the district south of the river 20 April to 10 July. This

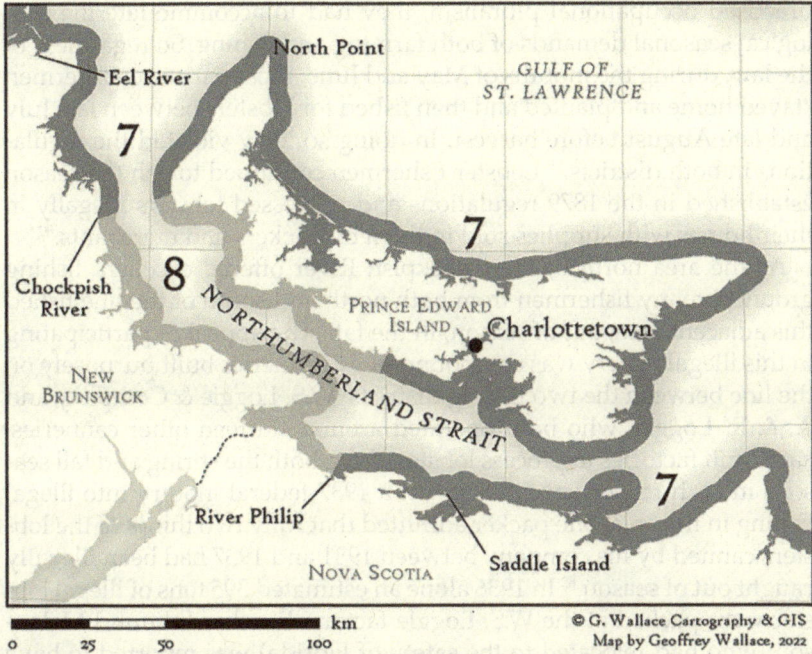

Figure 15. Map of Northumberland Strait, showing Chockpish River boundary, Saddle Island, and the boundaries of Lobster Fishing Districts 7 and 8 in the 1930s. @ G. Wallace Cartography & GIS

division was intended to consider ice patterns and water temperatures and permitted legal fishing in both areas for June and the first part of July.[76]

Even veteran fishery inspector William Wakeham – typically moderate in his assessment of the lobster fishery crisis – voiced concern in his 1910 report about the potential harm of summer fishing in the Northumberland Strait. He argued that the Strait's warm July water compromised lobster reproduction, and the quality of the lobster processed. Wakeham proposed a distinct fall fishery tailored to the unique ecological conditions of the Strait, but this was at odds with his general pragmatic inclination to encourage laws that could be enforced.[77]

In September 1918 the so-called fall fishery was introduced. The section of water south of the Chockpish River permitted fishing between 10 August and 15 October while the water north of the line continued to pursue a spring fishery.[78] This change created two sources of conflict. Since many of the households on both sides of the Chockpish Line practised occupational pluralism, they had to accommodate the ecological seasonal demands of both farming and fishing. So regardless of the law, during the months of May and June, most part-time fishermen stayed home and planted and then fished for lobsters between late July and late August before harvest. In doing so, they violated the regulations in both districts.[79] Lobster fishermen continued to fish the season established in the 1879 regulations and processed lobsters illegally in their homes with supplies sold to them by packers and merchants.[80]

As the area north of the Chockpish River offered excellent fishing grounds, many fishermen from both north and south of the line fished this adjacent area, out of season, in the fall. The appeal of participating in this illegal fishery was compounded by canneries built purposely on the line between the two fishing districts. W.S. Loggie & Company and A. & R. Loggie, who both operated around thirteen other canneries, each built factories to process lobsters from both the spring and fall seasons in each respective district.[81] At a 1937 federal inquiry into illegal fishing in the area, one packer admitted that fully two thirds of the lobsters canned by his company between 1931 and 1937 had been illegally caught out of season.[82] In 1936 alone an estimated 395 tons of illegal lobster were packed at the W.S. Loggie factory. Poacher Edmund Melanson (who had relocated to the safety of Florida) was reported to have bought 7.5 tons of Prince Edward Island lobster during the fall 1936 season and sold it in New Brunswick to A. & R. Loggie at Chockpish under the cover of night using coded flashing lights. Arthur Bourque, an employee of Émile Paturel, located in Shediac, estimated that more than 300 tons of lobsters were moved from District 7 to District 8 during the close season in 1936.[83]

The temptation of poaching lobsters was "irresistible" in the 1930s as the tough economic conditions expanded the number of fishermen willing to break the law. Although the presiding chair at the 1937 commission proceedings exhibited some sympathy for fishermen on the water as a "no man's land," he identified the active complicity of packers and buyers in violating the law. In 1937, in response to the commission's findings, the northern division line between Districts 7 and 8 was moved farther north to unify the New Brunswick Acadian coast within the Northumberland Strait area to reduce out of season poaching.[84]

The desire for practical and enforceable regions was also expressed at the community level. In 1902 an experienced Nova Scotia local fishery overseer wrote to the local Member of Parliament emphasizing that it was "very very important that the regulations for the three Western Counties be uniform." Differences were "unworkable" and "ensure violations and leave the officers powerless to detect or convict, the lines on the water being indefinable and imaginary."[85] The same overseer repeated the phrase "imaginary line" two years later in a letter to the Dominion Commissioner of Fisheries. Faced with enforcing new legal season divisions, he pointed out that "different regulations for adjoining sections are unworkable as each section will enjoy the others' privileges by simply crossing an imaginary line in the water." The same abstraction was a source of frustration for the chair of the 1937 commission, who concluded that "the exact position of the line is a somewhat nebulous conjecture or at least very uncertain."[86]

That bureaucrats and those charged with interpreting the law might raise concerns of policing imaginary lines is perhaps not surprising. By the late 1930s, however, the phrase "imaginary line" occasionally slipped into the official regulations and Orders of Privy Council. In November 1939 the inner Bay of Fundy region was adjusted so that it extended from the Annapolis Basin and the Digby Gut "as well as the entrance thereto, that is east of an imaginary straight line drawn from the southwest side of Bear Island to the fairway buoy about one-half mile off Point Prim, Digby county."[87] Here, the "imaginary line" was reified in law.

A Lobster War

On the southern side of Districts 7 and 8 another conflict emerged that brought together both traditional understandings of territory, formal fisheries regulations, and the use of formal and informal power. Cumberland County's Saddle Island's "lobster war" reminds us that state power cannot be entirely dismissed, as ordinary fishermen appealed to it to support their notion of justice. As already noted, the Department of Marine and Fisheries archives contains many examples of letters and petitions demanding the coercive power of the state to protect exclusive use of traditional fishing grounds.[88] It is easy to point to many examples of the failure of policing regulation and limited state capacity, yet individual fishermen, communities, and packers at times turned to the state for help.

In the 1930s the lobster season around Saddle Island, located in District 7, lasted only eight weeks from the first of May to the first of July. Its proximity to River Philip, the eastern border of Fishing District 8, meant that the area was attractive to fishermen from that area who had a fall open season. The Saddle Island lobster war came about as conflicts around territory and resource access coincided with depressed market conditions and high unemployment. The Depression had an immediate impact on consumer demand for canned lobster as a luxury good.

A Halifax newspaper argued, "The spending power of consumers has been reduced and the trade insists that unless lobsters are cheapened at the same ratio as the price of other foodstuffs are declining, the demand will be of less than formerly." Prices for canned lobster tumbled, as its reputation for poor quality had to contend with popular new consumer alternatives such as Japanese canned crab.[89]

The people living in the Malagash Point and Saddle Island area were Protestants of Anglo-Celtic background who made their living through farming, seasonal lumbering, a local salt mine, and the lobster fishery.[90] The May and June lobster fishing season meant that this was not a business that could be easily accommodated with farming, and as a result, fishing labour was imported from elsewhere. Before the 1920s local canning factories recruited labour for both the boats and the factories from Nova Scotia's Eastern Shore, but by the 1930s, most packers relied on Acadian women and men from Cape Ball, Shediac, and Bouctouche, New Brunswick for factory work and fishing respectively.[91]

The Saddle Island cannery was the largest in the area, but it was only one of seven on the small Malagash peninsula. A cannery had existed on the island since the 1880s, but around 1923 a new building was constructed by the Portland, Maine-based firm Burnham and Morrill, and then, according to its current business strategy, it was leased by the season to a local entrepreneur. Smaller independent packers were tied to wholesale chains through credit relations and leases. These packers in turn either directly hired fishermen to fish for them or claimed their catch through credit obligations.[92]

In the early 1930s George Tuttle King, from nearby Tatamagouche, leased the Burnham and Morrill factory. King operated with a combination of company and "independent" fishermen who owned their own boats. He hired a mixture of Acadian fishermen from Cape Tormentine and Port Elgin in New Brunswick and Anglo-Celtic men from his home of Pugwash, all in Lobster Fishing District 8.[93]

George Langille was from Malagash Point and operated two lobster factories, one just to the west of Saddle Island and the other to the south within the Amet Sound. In 1933 it was estimated that he had about $10,000 worth of capital tied up in gear, 500 traps in the water, and employed thirty-five fishermen and forty-five hands in the factories. Each factory was relatively small. Although in 1933 he had twenty-four boats at the northern factory and thirteen at the other, by 1938 there were only six boats fishing out of the north factory.[94] Langille had hired Eastern Shore men for his boats in the 1920s but in the 1930s, he claimed that most of his fishermen came from the nearby Wallace and Malagash area.

In the early 1930s a bitter rivalry developed between non-resident King and the fishermen he brought in from District 8 to work his Saddle Island factory and local George Langille and his "local" District 7 men. Falling lobster prices spurred fishermen to be even more competitive in their claims to the best fishing grounds, and lobster prices were especially low in May each year, as the Northumberland Strait season overlapped with the end of the lobster-rich area of the South Shore and Bay of Fundy. Tension came to a head in May 1933 when reports of a "lobster war" reached the provincial capital. Certainly, the Halifax newspaper exaggerated when it reported "several hundred lobstermen ... engaged in open battle as to which fraction will reign supreme on the fishing grounds."[95] In describing the situation at Saddle Island, one fisherman recounted that Tuttle King had leased the island and had about ten boat fishing "its grounds from the shore" and when other boats encroached, the King men began cutting trap lines and destroying buoys. Another fisherman later explained there were no formal laws but "conventions held that you should limit your fishing grounds to the area of your factory, wharf or packing outfit." Still others, while acknowledging these rights, also held that waters were marine commons and belonged to the local community who fished according to particular practices. This meant that when Langille and his crew encountered even a couple of boats of Acadian French Catholic fishermen from New Brunswick setting their gear in a different configuration than what was done locally, they felt justified in cutting the lines of the outsiders.[96]

Langille also believed that King's men were encroaching on his customary territory on both the west and the south and so had been provoked. According to Langille, King had threatened to drive him off the coast and push "him until he didn't have a boot on his foot."[97] He certainly was under pressure as he acknowledged that his catch was about 40 per cent of what it had been in 1932 and that he might have to close one of his factories.[98] The competition and ill feelings between the two operators was mirrored in their respective crews, who tended to be consistent across different seasons. Rarely did a fisherman have a work history that included both Langille and King.[99]

In 1933 the trouble began with cutting the lines to traps. One of the traditional ways in which lobster fishermen defend what they consider "their" territory is to cut the ropes linking buoys and traps. This results in either lost gear or the expenditure of considerable effort and time in retrieving traps from the bottom. Trap cutting can escalate. First cuts were threats or warnings, but both sides held their ground. The Langille side acted first, but after a three- or four-week period, King's men retaliated by cutting "all the Langille lines they could."

Verbal threats were exchanged, and guns were displayed. After a particularly active Friday night of trap cutting in late May with at least one gun shot fired from Saddle Island at one of Langille's boats, the Royal Canadian Mounted Police – the authority responsible for local policing – intervened. The RCMP estimated that 800 traps had been lost on Waugh's Shoal that season, and it established a visible presence while the Department of Fisheries cruiser *Arleux* was dispatched to "keep the peace."[100] The situation was volatile, as most of the men carried guns in their boats because there was a bounty on seals and the fishermen could always use the extra cash.[101] In an attempt to deescalate the situation, the Department of Fisheries inspector from Halifax was "unable to say" if one shot had been directed especially at George Langille or if Langille's passing may have coincided with an attempt to shoot a seal.[102]

The rivalry between the two packers' crews for the prime fishing area solidified the idea that French-speaking lobster fishermen from New Brunswick, who had a fall season, should be kept out of the spring season to the east. The *Halifax Herald* reported that the tension was not new but "aggravated this year by the general conditions of unemployment." Langille cast himself as employing only local fishermen (this wasn't quite true, but all his men were English-speaking Nova Scotians) while the factory on Saddle Island "imported" a small group of men from New Brunswick. That early reports stated that the fishermen were from Quebec should immediately alert us that this was a situation exacerbated by largely imagined differences based on religion and language. Coverage in the *Halifax Chronicle* described it as a war between fishermen from Malagash and Wallace and outsiders from New Brunswick and Quebec, stating that "rifle shots were fired at the Nova Scotians" and that it was the "invaders from the north who cut traps."[103]

Langille responded not only with direct informal action – the destruction of fishing gear he maintained contravened local practices – but also with a call for formal regulation and enforcement from the federal government. Like other packers before him who had petitioned the government to extend or change legal seasons, Langille organized a petition to request that the Department of Fisheries restrict lobster licensees to fishing only one season per year. Fishermen from New Brunswick who lived in District 8 should not be permitted to relocate to fish the District 7 spring season. The petition was signed by most of the Wallace fishermen and was adopted by the government. As of November 1933, no fisherman was permitted to fish lobster in more than one district in any given year. Consequently, this regulation ensured that fishing for lobsters could never be more than a brief seasonal occupation.[104] The significance of any packer or fisherman asking to be regulated cannot be underestimated, but it is puzzling to imagine that anyone believed these rules could be enforced at the local level.

The new regulation did not solve the problem. For the next four years cutting lines continued during the lobster season, although many traps were set without buoys to hide their location.[105] A fisherman recounted that if you had laid 500 traps at the beginning of the season, you might land only 300 or 400 traps eight weeks later. Packer Tuttle King compensated his men for these losses. Although incidents no longer made the newspaper, the occasional shot was fired, and on at least one occasion rival crews threw bait at each other.[106]

In 1937 Tuttle King moved his operations to Antigonish County and his brother-in-law Clarence Kennedy obtained the Burnham and Morrill Saddle Island factory lease. This added to George Langille's sense of grievance, as he had approached Burnham and Morrill representatives several times to gain the lease himself.[107] Kennedy was from Rockley, west of Pugwash, almost on the District 7 and 8 line. He had been in the lobster business since the First World War and operated a District 7 spring cannery nearby at Cape Cliff, Fox Harbour, and a District 8 fall factory in St.-Louis-de-Kent in New Brunswick. Lobster was a family business as his wife ran the cookhouse, and both of his sons entered the industry.[108] For the 1938 season at Saddle Island, he employed eleven independent fishermen, with single men fishing about 300 traps while two-man boats tended between 400 and 650 pots. All together there were about 4,000 traps, with 3,100 belonging to his men and his brother-in-law Tuttle King. King placed the value of the buildings and property on Saddle Island at $25,000, while Burnham and Morrill claimed that the plant was worth only between $6,000 and $8,000.[109]

Among the men who began the 1938 season on Saddle Island was forty-two-year-old Charles Corkum, a native of East Chester, Lunenburg County, who had come to Cumberland County seven years earlier via the Eastern Shore. Corkum offered his occupation as labourer – not fisherman – and this reflected his precarious income. Although he owned a boat, rope, and traps, he also worked seasonally as a day labourer on farms and in lumber camps. A married man, he and his wife were living apart. Before Clarence Kennedy, Corkum had worked for Tuttle King and before that another packer, Ed Seaman. The genealogy of his work is relevant, as when Corkum worked for Seaman, he went into debt and King took over the debt on the condition that Corkum fished for him. When Kennedy took over the factory, Corkum's debt was transferred in turn. Although some factories operated with so-called independent fishermen, they were tied tightly to specific packers through ongoing credit obligations and were seldom free to sell their goods as they chose.

It was unusual then on the first of June 1938 – the middle of the season – that Corkum quit Kennedy and went to work for another packer at nearby Brule Point.[110] At the time, Corkum still owed Kennedy $130.86, and so he was not permitted to move his gear. In fact, even if he had been free to do so, it would have been very difficult for Corkum to move his personal things, as fellow fishermen had nailed his trunk and boots to the floor. Corkum later claimed that Kennedy's crew suspected him of "carrying news over to Langille's fishermen." They had not only fastened down his belongings but had also thrown bait at him and put water in his engine. He claimed he was "tormented and teased" as the Kennedy men made good on their threat that they were "going to chase me off the Island." At this point rival packer George Langille loaned Corkum the cash needed to pay his debts with 5 per cent interest and the promise that he would fish for Langille during the following season. When he settled his account with Kennedy, they "had words," as Kennedy was no doubt enraged that one of his fishermen was going over to his rival.[111]

Tension had been high between Kennedy and Langille's crews that spring with intense rope cutting on Waugh's Shoal. Sometime during the summer Langille approached a representative of the newly created Maritime Fishermen's Union to request Burnham and Morrill "get Kennedy off the Island." When a union representative responded that this local conflict had nothing to do with the union, Langille replied that he would "clean up the God Damn Strait." If he could not fish out there no one else would and that if his fishermen would not cut lines, next spring he would only hire fishermen who would.

According to Corkum, fellow lobster fisherman Percy Langille offered him "easy money," as George Langille would pay him $100 to "burn out" Saddle Island. On the evening of 15 September 1938, Charles Corkum stole two bottles of gasoline from a boat along the shore and set fire to Saddle Island's nine buildings, traps, rope, and boats.[112] In his trial he admitted to this charge and testified against Percy and George Langille, who were both charged with counselling arson.[113]

George Langille was cleared of the crime, but at least some local memory holds him responsible for paying Corkum. Circumstantial evidence such as an additional $25 cheque from George Langille to Corkum in December 1938 looked suspicious. Percy Langille was also found not guilty, but Corkum was sentenced to two years in a federal penitentiary and was released early for good behaviour and agreeing to join the army. The Saddle Island factory was rebuilt, and Kennedy continued to operate from there until the late 1950s when Malagash salt mines closed, and he moved his buildings and his operations to this mainland location. When the old mine warehouse used for storing traps burned down, arson was always suspected.[114]

The economic crisis of the 1930s encouraged many men without other economic opportunities to enter temporarily the lobster fishery, thereby increasing competition for scant resources. Informal strategies to keep outsiders away from what was perceived as a local community resource included cutting lines, firing shots at boats, harassment of disloyal fishermen, and ultimately arson. Parallel efforts of the state to end these "lobster wars" created opportunities for the state to increase its legitimacy among fishermen as fishermen themselves initiated new regulatory policies.

The legacy of the Saddle Island lobster war was more regulations that needed to be enforced – imperfectly. From November 1933 lobster fishing licences were supposed to be restricted to only one fishing district each year. By 1937 the first regulations restricting the movement of boats, traps, and gear to more than one fishing district in any year was introduced for District 8 and was extended elsewhere the following year.

The coercive power of the state had its limits, and fishermen and packers occasionally responded with their own use of violence or the destruction of property. Attempts by the state to encourage adherence to the laws and regulations through education, licensing, fines, jail, or the destruction or confiscation of property were uneven. The 1955 Department strategy of withholding licences from law-breaking fishermen after three convictions in a single season caused one fishery officer to observed that the sole consequence was "we only create one more man to watch full time."[115]

It is not surprising that there were hopes for self-regulation through education. W.F. Tidmarsh, general manager for the Portland Packing Company on Prince Edward Island, represented the "Canned Fish Section" of the Canadian Manufacturers' Association at a 1927 government commission. After forty years in the business, he concluded that lobster regulations could not be enforced and were "a source of irritation, not protection." He argued that rather than coercion "educational propaganda" would "do more to restore and protect the lobster fishery than can ever be accomplished by legislation."[116] As discussed in chapter 4, this was the very tactic being adopted by the Fisheries Branch at the time – with questionable success. After the Maclean Commission, the Department of Marine and Fisheries began making grants to the Extension Department of St. Francis Xavier University to promote conservation education among the fishermen through community action.[117]

In the post-war years, Fisheries ran educational campaigns—timed to coincide with local lobster seasons—using newspaper advertisements to promote the image of honest fishermen working toward their own material benefit by embracing conservation. The campaigns emphasized economic prosperity and the promise of the post-war good life. Ads such as "Your pot of gold can be a lobster pot" with a rainbow ending in a lobster trap reflected the department's preoccupation with income improvement and stabilization.[118] This was a change in tactic from the 1944 advertisement with information on lobster licences where in a didactic fashion there was an entire page explaining the consequences of destroying berried lobsters and how "YOU as a fisherman would suffer."[119] By 1956 the message was "good catches, bigger profits and more security" as "Conservation means bigger profits for you."[120] One advertisement in the campaign made reference to the community rather than individual profit – but it was a community where individuals benefited. Here, when lobster fishermen had a good season through conservation, "the whole community feels his success – the grocery store, the dress shop, the appliance store, the doctor, the dentist, the insurance agent and many others, all benefit. A poor catch means poor business – hard times."[121] The 1964 Department of Fisheries poster "It's your business to Protect your future" was a blatant appeal to self-interest – a way of approaching the problem of enforcing laws and regulations that may have been effective but helped to reframe the way access to the marine resource was seen as an individual rather than community issue. This was reinforced by the parallel suggestion that only full-time lobster fishermen be eligible for licences, as they would be more responsible conservationists and solve the issue of enforcement through self-regulation.[122]

The strategy was convincing individual fishermen, not communities or packers, that it was in their interest, as male breadwinners, to participate in regulations that sought to conserve the lobster stock and stabilize prices. Adult educational campaigns through film increased in the postwar period, such as the 1957 National Film Board production "The Trap Thief"/"Le pilleur de cage."[123] Set in a Cape Breton fishing community and filmed in Main-à-Dieu, the English version was broadcast on CBC's television series "Perspective." It featured an actual fishery officer, Aubrey MacKinnon of Louisbourg, and a real patrol boat, *Sabella*, under command of Captain L.C. King.[124] The thirty-minute fictionalized story pits the decent fishery officer (supported by the local priest) and conservation-minded honest fishermen against a young

Figure 16. Department of Fisheries advertisement, *Antigonish Casket*, 10 May 1956.

ne'er do well who not only kept "shorts" and "spawn" lobsters but also stole from his neighbours' traps. By the end of the story, the offender is brought to formal and informal justice, both before the courts and in the community. The prospect of a fine was described as the "least of it. He's finished around here."

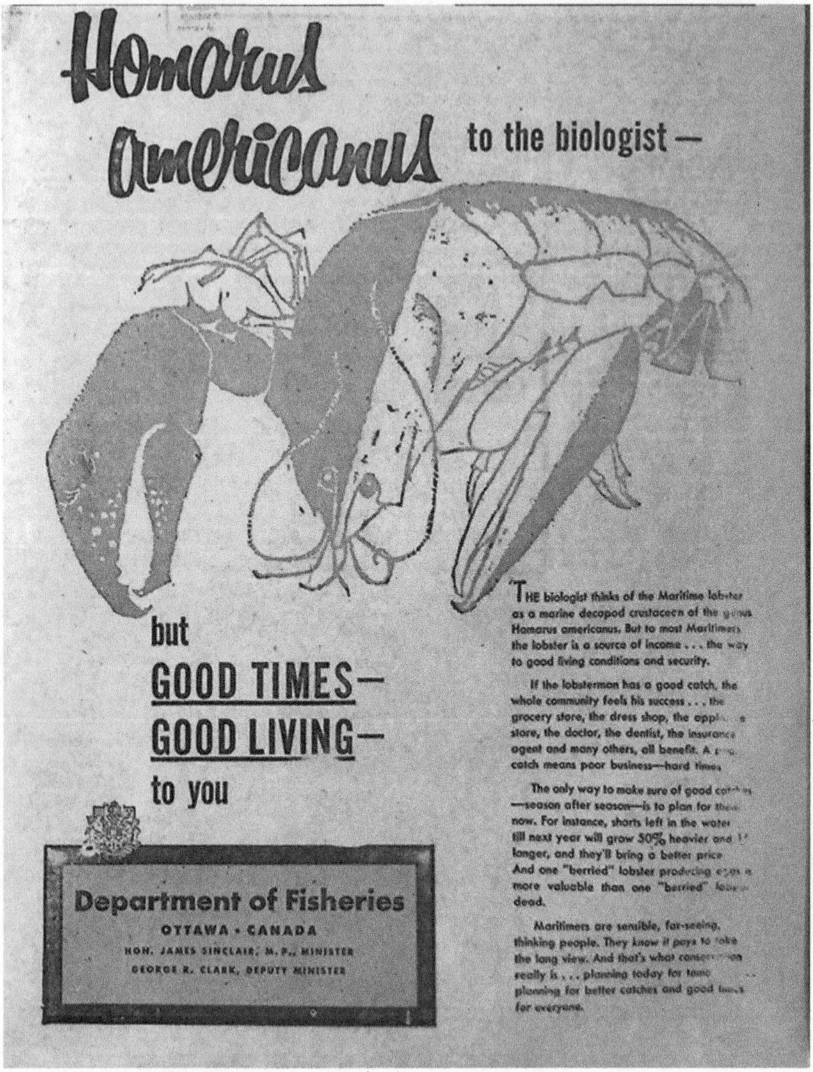

Figure 17. Department of Fisheries advertisement, *Antigonish Casket*, 31 May 1956, "If the lobsterman has a good catch, the whole community feels the success ... the grocery store, the dress shop, the appliance store, the doctor, the dentist, the insurance agent and many others, all benefit. A poor catch means poor business – hard times." "Maritimers are sensible, far-seeing, thinking people. They know it pays to take the long view. And that's what conservation really is ... planning today for tomorrow ... planning for better catches and good times for everyone."

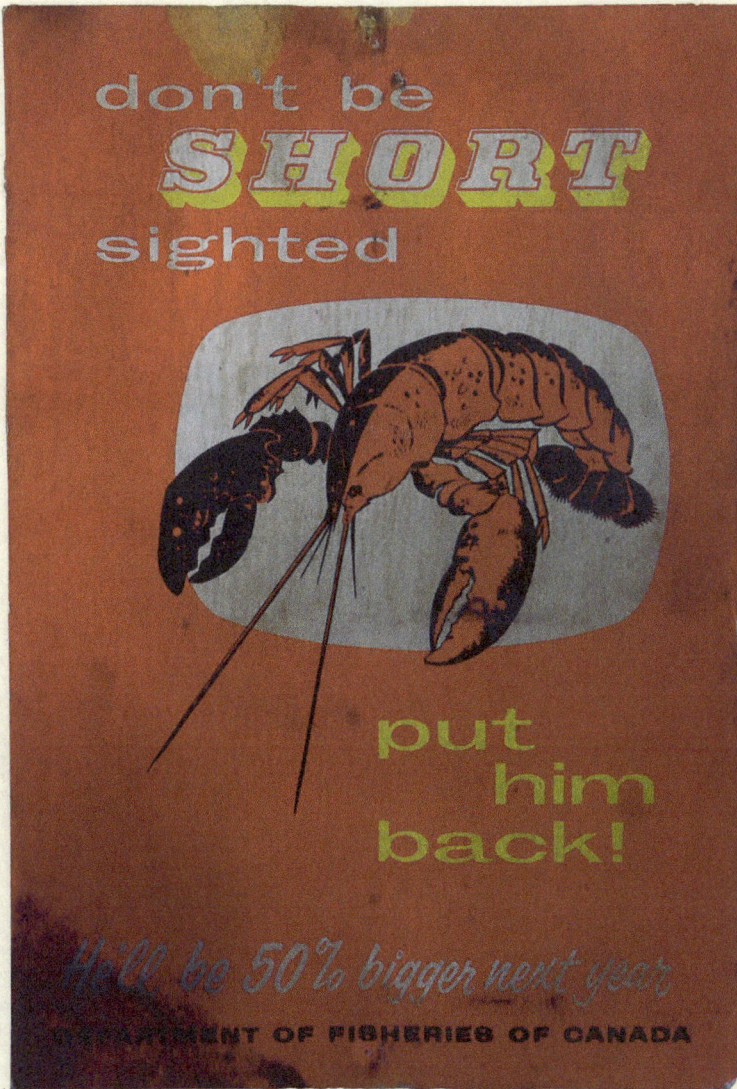

Figure 18. The Department of Fisheries distributed conservation posters to fisheries officers for local display. Like the newspaper advertisements, these posters emphasized the individual economic benefits of adopting conservation measures, such as throwing back lobsters that were undersized or that carried eggs.

Harley Clattenburg Collection in possession of Heather and John Stevens. With permission.

Figure 19. This poster presents lobster fishing as a business that stood to benefit from the enforcement of conservation regulations, which are touted as essential for both protecting and growing the industry.

Harley Clattenburg Collection in possession of Heather and John Stevens. With permission.

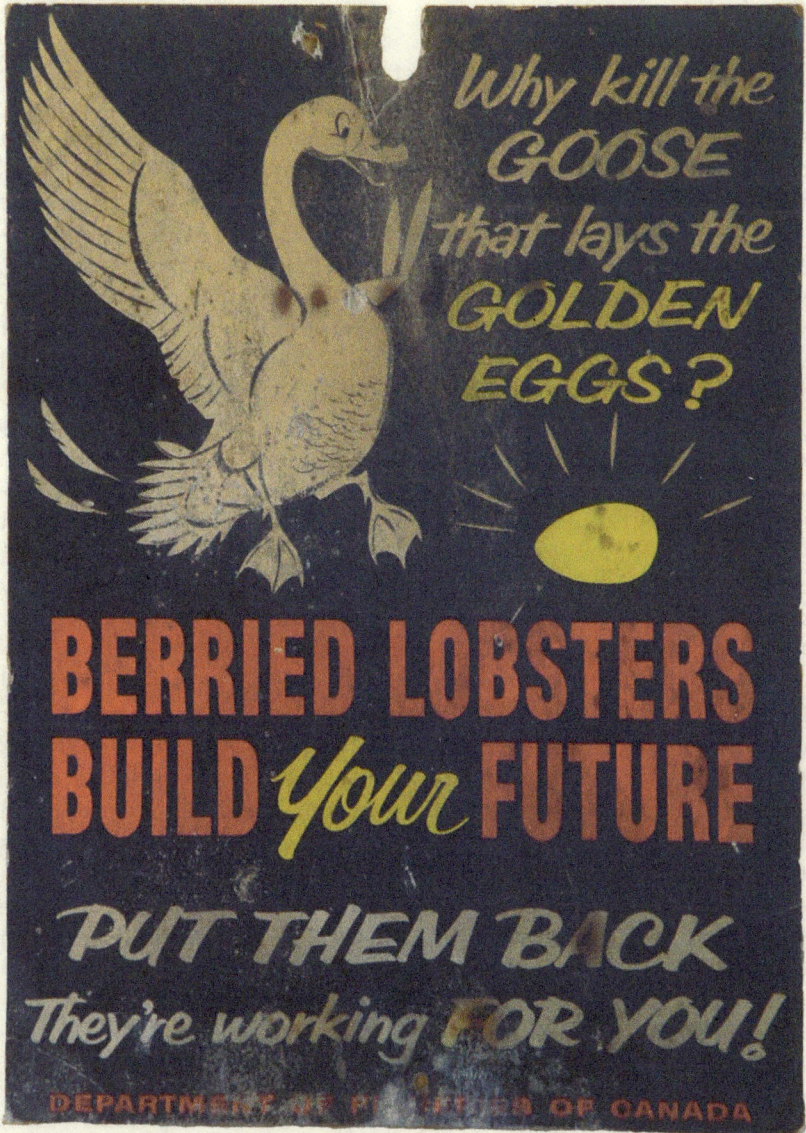

Figure 20. In this poster, female lobsters carrying eggs – known as "berried lobsters" – are depicted as a golden goose for future prosperity. By following regulations and "putting them back," fishermen would have a prosperous future.

Harley Clattenburg Collection in possession of Heather and John Stevens. With permission.

Conclusion

There is remarkable continuity between the 1890s and the 1960s in how the enforcement of formal laws and regulations exposed the limits of the federal state. Widespread illegal fishing persisted – not only because it offered an important source of cash but because it was widely supported or tacitly accepted in many communities. At specific times and places, the fisheries department lacked both the capacity and the authority to impose order – whether through physical force, judicial means, or moral persuasion. Clearly, not all canners and fishermen accepted the legitimacy of regulations. When they did comply, it was often because the rules aligned with their own interests or local strategies. When laws conflicted with those strategies, communities, packers, or individual fishermen paid no mind and fished. They also used or threatened physical force and the destruction of property to achieve their own understanding of justice and the right to fish lobster. Interestingly, informal practices turned to the state when it suited them – revealing how formal and informal systems were interwoven, much like the relationship between legal and illicit canners. Regulatory conflicts were usually most obvious at the "imaginary lines" drawn by the state that set open and close seasons.

The focus of state regulation shifted from fishermen to canners and back to fishermen, and with this change came different strategies for compliance. Over time the state increased its coercive power through technology and policing but also adopted new approaches to achieve social or community compliance. In particular, after the Second World War, Fisheries educational propaganda highlighted individual benefits that might be gained through community resources. As we will see in chapter 7, the blatant appeal to individual self-interest helped lay the groundwork for what by 1970 would be a limited access fishery. But even after this dramatic change, there was continuity in local community beliefs about who had the right to fish where and when. These local practices were shaped in part by the very state regulations they resisted, blending official rule with informal enforcement in ways that reflected deep-rooted understandings of entitlement and place.

Chapter Six

The Politics of Lobster and Catching Votes, 1871–1939

In 1921 *Maclean's* magazine attempted to educate its largely Central Canadian readership about the importance of the lobster industry with the bold statement, "Lobster is another name for politics on the Atlantic coast."[1] The politics of lobster took both formal and informal forms, and even in its informal expressions, the state was seldom completely absent. Political parties campaigned for the fishermen's vote or the canner's purse, dispensed patronage appointments, orchestrated political interference in the judicial process, and negotiated with lobster-related interest groups who organized petitions and letter-writing campaigns as mechanisms of political change. At the community level, groups of fishermen came together to promote their own interests or establish cooperatives, becoming recognized institutionalized stakeholders and providing feedback in the development of government regulations. Those in the packing sector used informal cartels and recognized associations such as the Canned Fish Section of the Canadian Manufacturers' Association to advance their interests through formal public appeals and behind-the-scenes lobbying. While the previous chapter dealt with direct forms of action that clearly had political intent, as they were aimed at bringing about change and addressed imbalances in power and the law, the lobster fishery was also emmeshed in the most traditional definitions of politics. While lobster politics remained significant, the direct electoral influence of both canners and fishermen gradually declined, mirroring the diminishing political weight of the Maritimes within Canada and its reduced impact on national political outcomes.

Most federal constituencies in the Maritimes were ridings where lobster played an important role in the economy. There were political costs and benefits to regulating the fishery: enforcing laws, catering to fishermen, pleasing the packers, and distributing patronage all carried electoral consequences. This was especially true in Prince Edward Island

and most Nova Scotia ridings – excluding Hants, and the relatively minor lobster fisheries of Annapolis and Kings counties. In New Brunswick ridings, all coastal constituencies had significant lobster fisheries, with the notable exception of Albert County.

The considerable number of federal ridings where lobster was fished commercially shaped federal lobster policy in two separate ways. In the first place, it amplified the importance of the lobster industry well beyond its actual significance to the overall Canadian economy. In the big picture, fishing – in all its forms – has played a very minor role in determining the Canadian gross national product ranging from 0.8 per cent in 1870 to almost 1.6 per cent in 1896 and about 0.6 per cent in 1926.[2] This small percentage includes lobster and the then abundant West Coast salmon and Atlantic groundfish. The politics of lobster have never been about wealth as much as employment and voters. This is apparent in the pivotal 1896 election, which saw the defeat of the Conservatives and the success of Wilfrid Laurier's Liberals. Here, for example, lobster mattered in thirty seats of the 213 national ridings. This meant that the lobster fishery was significant for over 14 per cent of all electoral contests. As the proportional weight of the region's seats in the House of Commons declined, the Maritimes had less influence on the electoral success of national political parties. Correspondingly, lobster regulations became less of a visible political issue. By 1965 only twenty of the then 265 seats in the House of Commons were in lobster ridings. With only 7.5 per cent of all elected members, lobster politics held less sway in the outcome of federal elections. It is hard not to see that the emergence of non-partisan political organizations such as local cooperatives coincided with the diminished electoral political importance of the region.

Secondly, the highly politicized nature of lobster regulations was connected to the very small margins of victory for either the Conservatives or the Liberals in a remarkable number of these federal "lobster ridings."[3] Lobster policy gained and lost votes, and it could make a difference in the formation of a government or its defeat. In the 1887 election, for example, in thirteen lobster ridings, the winner claimed a plurality of less than 3 per cent of the popular vote. In the 1896 election, this number rose to eighteen seats. With the return of the Conservatives in 1911, the margin of victory to either party in eleven lobster ridings was extremely close. In general, until after the First World War, most lobster ridings were swing ridings and had the potential to elect either party. Even during the fractious 1917 wartime election, the results were much closer than the distribution of seats suggest. Although fourteen of the twenty-two lobster riding seats went to the Unionists, if

Table 2. Seats in the House of Commons and Percentage of Regional Seats in House of Commons, 1871–1966

	Nova Scotia	NB	PEI	Nfld	Total Seats	% of seats in the House of Commons
1871	19	15			185	18.4
1872	21	16			200	18.5
1873	21	16	6		206	20.9
1882	21	16	6		211	20.4
1887	21	16	6		215	20
1892	20	14	5		213	18.3
1903	18	13	4		214	16.4
1907	18	13	4		221	15.8
1914	16	11	3		234	12.8
1915	16	11	4		235	13.1
1924	14	11	4		245	11.8
1933	12	10	4		245	10.6
1947	13	10	4		255	10.6
1949	13	10	4	7	262	13
1952	12	10	4	7	265	12.5
1966	11	10	4	7	264	12.1

Source: https://lop.parl.ca/sites/ParlInfo/default/en_CA/ElectionsRidings/Elections

the total number of votes was amalgamated, the Unionist candidates only received 49.78 per cent of the total votes cast.[4] The early 1920s is associated with the failed Maritime Rights movement, a business-led campaign linked to the Conservative Party aimed at federal domestic policies that hurt the region's economy. Lobster as an export product was almost entirely outside expressed concerns except for complaints about policing regulations. Nonetheless the Conservatives swept sixteen of the twenty-two "lobster ridings" in the 1925 election but with only 53 per cent of the overall vote.

The significant number of seats at stake and the small margins of victory for either the Conservative or the Liberal parties help explain the prominence of lobster policy in local politics, especially before the First World War. The 1921 article in *Maclean's* magazine that began this chapter noted that 40,000 or 50,000 votes were connected to the industry, and "this formidable voting power is sufficient to sway about twenty-five seats in the Dominion House and a majority of seats in the various

Maritime Houses of Assembly."[5] The influence was compounded further by the number of provincial and federal politicians who were involved in the lobster industry.

Direct and indirect political intervention permeated almost every aspect of the creation and policing of lobster policy. This meddling took many forms that ranged from patronage appointments and the awarding of government contracts to direct interference into the operation of the fisheries department and attempts to leverage the vote in favour of a specific political party. New Brunswick and Nova Scotia canner, and a Liberal politician himself, Fred Magee concluded in 1937 that the lobster fishery had been "bedevilled with politics since 1896" and that all guardian positions remained political appointments and therefore subject to political interference.[6] This was very much in keeping with journalist Charles C. Jenkins's conclusion in 1921 that "political intrigue" connected to lobster votes meant that "party men rather than practical ones" determined policy and there was lax regulation and haphazard enforcement of the fishery.[7]

The use of patronage or "the spoils system" in providing temporary but very valuable employment has already been mentioned, but it is difficult not to belabour this point. It ran roughshod over the Department of Marine and Fisheries' or Department of Fisheries' attempts at professionalization. In the 1920s West Prince district, the appointment of local guardians was made after recommendations by a local patronage committee composed of individuals who themselves poached lobsters. The result, according to Justice A.T. LeBlanc, who chaired an investigation, was that these political appointments "protected the fishermen instead of the lobsters." In other communities, the patronage committee making appointments were composed of the same packers the fishery officers they named were supposed to police.[8] Appointments were awarded to failed and defeated candidates at either the provincial or federal level. Even as the civil service was undergoing reforms to make it less bound to patronage, Ward Fisher, the defeated Conservative candidate in Nova Scotia's 1911 provincial election, was appointed as Fisheries Inspector for Nova Scotia District 3.[9]

W.F. Tidmarsh, who was responsible for the Portland Packing Companies' operations on Prince Edward Island from 1885 to 1935, harshly concluded in 1909 that local fishery wardens did nothing, as they had received their appointments as compensation "for political services." Tidmarsh proposed the removal of policing from the political sphere completely with the creation of a specialized force in which appointments were not made by politicians and, like the RCMP, officers could not serve in their home provinces, thereby distancing them from family and community pressure.[10] The same issue was raised ten years later by R.H. Williams, of the leading international wholesalers Roberts, Simpson and Company, who wrote in the Halifax *Morning Chronicle* in 1918 that Fishery Overseers were "one of the fruits of our Canadian style of politics ... He's usually strong Grit and strong Tory and in politics his strength is spent so that there is little left for anything else unless he believes it will influence the next election and solidify his position in the future." Their trifling salary at $30 or $40 a year meant the government could expect little actual work from them.[11]

As their positions in local communities was often more political than regulatory, those who performed their tasks faced consequences. This was the case for a fishery officer who was a former lobster fisherman, cannery manager for W.S. Loggie Company, and for seven years in business for himself. He reported to the 1920s Maclean Commission that after five years as an officer he had been dismissed from his post for "being too strict with the illegal fishermen."[12]

Appointments at the level of inspector and above were permanent positions, part of the regular civil service, and therefore supposedly above direct political involvement. Yet, in 1905 canner Michael Joseph Neville complained to William Roche, a Liberal Member of Parliament for Halifax, that the Fisheries Inspector for eastern mainland Nova Scotia Robert Hockin, who had been appointed by the previous Conservative government and was "not a friend of ours politically, and it does seem to me that in some instances he shows the cloven foot." Neville not so subtly hinted that there were Liberal "friends" available who would be pleased to have the position. This political lobbying did not meet with success, and Hockin kept his position from 1890 for nearly thirty years.[13]

Yet there clearly were assumptions being made about the political affiliations of even senior members of the Department. In 1911 Henry Le Vatte, a Cape Breton Liberal who had served on the 1898 Royal Commission, wrote to W.A. Found, Superintendent of Fisheries from his position as a fishery officer in Louisbourg (a position he would lose with the election of the Conservatives that year). His claim that many of his fellow officers failed to enforce lobster regulations because they wanted to keep their "$55 per annum plus expenses" was not surprising, but his political candour in addressing the senior civil servant responsible for fisheries operations was. Six months before the federal election, he relayed to Found that there was local "antipathy to the changed Lobster regulations" that would "certainly be a very excellent canvas for our political opponents."[14] Here Le Vatte's assumption that a senior civil servant shared his political objectives is striking.

Direct political interference related to hiring continued into the 1950s and could both protect or disadvantage certain packers or political constituencies. In 1947 a Department of Fisheries official wrote to Reverend Moses Coady, one of the founders of the Antigonish movement and director of the St. Francis Xavier University Extension Department, conveying that the local Member of Parliament for Antigonish-Guysborough had requested that a Catholic be named as the new local fishery inspector. Coady was asked to confirm "the religious adherence" of proposed candidates. All the names provided had political qualifications rather than any connection to the fishery, and Coady's response noted the Catholics but added he would have preferred names "from rural communities and fellows in sympathy with the fishermen's Co-operative movement."[15]

The close relationship between patronage appointments and local politicians was one of the most obvious ways that politicians interfered with the operations of the Fisheries Department. In some cases, politicians protected their political appointees against charges of negligence. In 1901, a local fisheries officer in River John, Nova Scotia, was dismissed after he lost five hundred government lobster stamps intended to be used to designate legally packed lobster. The loss of so many stamps allowed illicitly packed lobster to enter the legal stream. After a complaint from the local Member of Parliament, the Deputy Minister of Marine and Fisheries reinstated the employee as a "misunderstanding."[16]

Robert N. Venning, as the civil servant in charge of the administration of the Fisheries branch between 1895 and 1911, made efforts to keep politics out of the enforcement of lobster regulations. However, local fisheries officers often communicated directly with politicians who had secured their appointments, rather than the Department itself. As patronage appointments, fishery overseers relied on the goodwill of elected officials, and this allegiance could put them at odds with department bureaucrats. In 1898 Abram M. Hatfield, the Yarmouth area Fishery Overseer (whose son, Paul Hatfield, would later serve as the Liberal MP for the area) was reprimanded for bypassing department channels. He had written directly to the influential W.S. Fielding (who was not even his local Member of Parliament) regarding an enforcement dispute with the Department, in an attempt to circumvent its authority.[17] Both Fielding and Frederick Borden, in particular, involved themselves in almost every decision that touched Nova Scotia.[18] As Minister of Finance and representative for the lobster district of Shelburne and Queens from 1896 to 1911 (and again 1917–25), Fielding was especially active in placing his interests before Fisheries by letter, memo, and telegram. Almost ten years later, in 1907, Venning had to reprimand another overseer in Canning, Nova Scotia, after he wrote directly to his local elected Member Sir Frederick Borden, Minister of Militia and Defence. The Superintendent of Fisheries noted that it was "highly improper" for any fisheries employee to communicate directly with an elected politician about his duty as a fishery officer. In this case, the overseer provided incorrect information about lobster shipping that further embarrassed the department.[19] A year earlier, in an internal memo, the Deputy Superintendent of Fisheries had complained to his superior, Venning, of the "political meddling" of Borden. When defending a local resident accused of handling lobsters out of season, Borden demanded an immediate investigation and that "his representations be taken as complaint."[20]

This type of intervention into department operations was another way in which politicians shaped practices. In late April 1907, William Roche appeared again and requested that the Department "move slowly" on the collection of fines for constituents successfully prosecuted for breaking lobster regulations.[21] This bid coincided with the campaign period for the June Nova Scotia general election. Individual Members of Parliament also petitioned on behalf of their constituents. In 1903 Fletcher Wade, the Liberal Member for Annapolis, was successful in getting the local inspector "to refrain" from collecting the fine from a local fisherman caught with lobsters in the close season.[22] Examples of political interference were raised again in the 1936–8 Inquiry into Illegal Lobster Fishing and Canning (and Illegal Smelt Fishing) in Lobster Fishing Districts Numbers 7 and 8 when elected local federal politicians intervened on behalf of friends. One witness recounted the example from around 1920 of a Port Hill, PEI, lobster fisherman who was found guilty and sentenced to a fine of $300 or six months in jail. The fisherman took the jail term, and his subsequent release after only six weeks served was attributed to "political pull."[23]

At times it is impossible to differentiate between political and judicial interference in the handling of prosecutions. Just before the First World War, a local overseer recommended that a case related to illegal canning be tried in nearby Sackville, New Brunswick, as it was difficult to get a magistrate to try a case in Port Elgin "owing to the influence of these men."[24] Interference could take place at the highest level. In 1891 the Deputy Minister of Fisheries wrote the Deputy Minister of Justice conveying the request of the Conservative Member of Parliament for Shelburne County that a $100 fine imposed on the Wood Harbour Packing Company be reduced to $20.[25]

While interference in the policy or judicial process usually took place at the individual level, on at least one occasion Members of Parliament coordinated their actions in an unsuccessful attempt to strike at the very centre of conservation policy. Three months after the 1896 election, a joint letter from all of Nova Scotia's Liberal Members of Parliament to the Minister of Marine and Fisheries declared that they were against close lobster seasons and that "fishermen ought not to be hampered by vexatious legislation." This was a remarkable position for a fishery in need of protection, and most members' positions were softened with education during their time in office. The letter also made the point that those responsible for the enforcement of lobster regulations could not be trusted, as "The present officials are almost all Conservative, and they have made the Department a political machine in the interests of the Tory party."[26] Liberals used the next fifteen years to do the same thing.

Almost every aspect of the Department of Marine and Fisheries was political. Local politicians regularly lobbied for their political friends. W.S. Fielding forwarded correspondence from local Liberals recommending the appointment of a specific man to the 1898 Royal Commission on the Lobster Industry (complete with his thirty-year Liberal pedigree without ever receiving "any consideration for his long and valuable support of the party.")[27] He also petitioned the Minister in support of "friends" who wanted the department to charter their boats for seasonal regulation enforcement. Lists of "friends" hotels were also provided for the many commissions and fisheries employees deferred to politicians on these issues as a matter of course.[28] When the use of patronage appointments for fisheries officers became less blatant, partisan politics adopted other forms. For example, in the postwar period, local Nova Scotian communities worked through the Liberal Party to get new public wharves constructed.[29] Progressive Conservatives also adopted this strategy. The 1958 election of former lobster buyer Conservative Felton Légère to represent Shelburne-Yarmouth-Clare saw a local boom in the construction and repair of public wharves.[30] These wharves were (and are) essential to the lobster fishery, as they permitted resource access to those who did not own ocean frontage.

Political special consideration was supposed to be discrete, and on occasion it was. In 1903 the Minister of Marine and Fisheries acknowledged that, although he had given "no publicity of the fact," the local inspector at Pleasant Bay, Inverness County, had been told not to "interfere" with lobster operations, thereby providing the equivalent of a private and secret extension to the local open season.[31] The same year, W.S. Fielding, among other federal Liberal Members of Parliament from Nova Scotia, lobbied for the introduction of breeding pounds to be located in their particular ridings. A compromise was reached with a pilot project, with one pound in Cape Breton belonging to an individual with Liberal connections.[32]

Expectations of special treatment extended to lobster fishermen. Individual fishermen writing to the Minister of Marine and Fisheries stated their political affinity to advance their case. A 1910 letter to Minister of Marine and Fisheries Louis-Philippe Brodeur closed with the salutation, "Yours truly Liberal supporter Isaac Wambolt."[33] Others such as James Smith of Canso, Nova Scotia, offered more disingenuous closures, such as "I have been a Liberal since the days of the late Joseph Howe but I ask no favour on that account."[34]

One of the factors encouraging special treatment between politicians and packers was how closely they were entwined. A remarkable number of provincial and federal politicians were in the lobster business. Twenty-six members of the 325 Prince Edward Island legislators who served between 1873 and 1970 were identified specifically with the lobster industry in their legislative biographies.[35] This is an under-representation, as few Island merchants in the late nineteenth and early twentieth centuries were not somehow connected to the lobster trade. Among those provincial politicians not identified specifically with lobster in their biographies was Conservative MLA and later senator Samuel Prowse. Prowse operated at least two lobster canning factories and even manufactured his own tins.[36] Conservative Member of Parliament in the 1870s Daniel Davies was identified as a merchant and shipbuilder, but an 1891 report in the *Daily Patriot* described his considerable losses in a fire that included a lobster factory at Murray Harbour South.[37] Other politicians were linked to lobster. Among the most prominent packers in western Prince Edward Island in the 1880s was Acadian Joseph-Octave Arsenault who served as both a Liberal and Conservative in the Provincial House and was an important leader in the Acadian community. Lobster interests extended to the height of Prince Edward Island politics. Liberal Premier (and later federal politician) Donald Farquharson owned a lobster factory at Canoe Cove among his many interests.[38] Island merchants drawn to federal politics were also steeped in the lobster trade.

In Nova Scotia merchant-politicians came from more diversified sectors, but in the late nineteenth century all packers but three on the Atlantic coast were said to be identified with the Liberal Party.[39] This might hardly be surprising in a province where between 1867 and 1956 the Liberals held provincial power for seventy-six of the eighty-nine years, often with literally less than a handful (fewer than five!) members in opposition. A directory of provincial legislators before 1970 identified only five members connected with the lobster industry, but this number is also an under-enumeration. The omission of lobster from Moses Nickerson's career alerts us to the limitations of this source. Nickerson was by far the most prominent Nova Scotian politician identified with the lobster industry.

New Brunswick provincial politicians lack a specialized historical database, but the *Dictionary of Canadian Biography* offers examples such as Liberal Member of the Legislative Assembly Robert Young, who had canneries in Caraquet and Little Shippegan. New Brunswick provincial and federal Conservative politician Joseph L. Black owned and operated a lobster factory at Cape Tormentine in the 1890s. Olivier Melanson was a municipal and Conservative provincial politician and around the twentieth century was involved in at least a dozen factories.[40] No Acadian politician was associated with greater wealth from lobster. Pioneer New Brunswick packer Henry O'Leary opened his first cannery in 1865 and by 1895, in addition to his Prince Edward Island holdings, owned more than thirty canneries in New Brunswick alone and served as a Liberal in that provincial house.[41] His son, Richard O'Leary, a Conservative, was also a lobster packer, among other activities, in New Brunswick and founded O'Leary and Lee in Nova Scotia, a wholesaling company which became one of the largest distributors of tinned lobster in the world. Richard O'Leary was not an elected politician but served on the 1912 federal Commission on Shellfish and has been described as "the acknowledged leader of the Conservatives in Kent County."[42] Federal Liberal Member of Parliament for Northumberland (later a senator and New Brunswick's lieutenant governor) Jabez Bunting Snowball was primarily in lumber but diversified into lobster canning on Lamèque and Miscou Islands. His firm shipped over a hundred tons of canned lobster from Chatham in 1879, or about 70 per cent of all lobster exported from that port.[43] Of note was William Stuart Loggie, a Liberal MLA, MP, and mayor of Chatham. He headed W.S. Loggie Company, which focused on fish and the canning and shipping of lobsters and blueberries. In the early twentieth century Loggie boasted as many as seven hundred employees on his payroll in the more than a dozen permanent factories he operated in Northumberland, Gloucester, and Kent, in addition to nearly forty smaller seasonal operations in Nova Scotia, Prince Edward Island, and Quebec.[44] Finally there is the example of Fred Magee, a Liberal MLA for Westmorland from 1917 until 1925 whose Mephisto Brand, first of Port Elgin, New Brunswick, and later Pictou, Nova Scotia, distributed canned lobsters around the world.[45]

This overlap between politicians and packers was an aspect of the nationalist politics (or at least anti-American sentiments) used by politicians of both parties and even the civil service as an effective strategy when most of the larger firms were owned south of the border. As previously noted, as early as 1872, when the commercial lobster fishery was just taking off, a fishery officer lamented the American presence in the Baie des Chaleurs and the Gaspésie, as the "wealth of our own waters seems to have fallen into the hands of American citizens."[46] Embracing a nascent Canadian nationalism, government employees in 1877 also raised concerns when branches of American firms used American labels and passed off Canadian lobster as a product of the United States. In 1889 information on American ownership and the presence of American employees was collected by the Fisheries Protective Service.[47] When this matter came before the 1898 Royal Commission on the Lobster Industry, the final report drew particular attention to what it called the "remarkable fact" that such a large part of the industry was owned by Americans headquartered in Portland, Maine. While three of the seven commissioners urged that no new foreign applications be granted for canning licences, the report more meekly concluded that it "would rather see Canadians favoured."[48]

The role of Americans in the canning industry was explicitly raised in a 1902 survey sent to members of the regional Liberal caucus. Members were queried how they felt about a clearly nationalist proposal that would give two years' notice before all licences were restricted to resident Canadians. Significantly, this potentially radical proposition, with no clear explanation, excluded the two largest American companies, Portland Packing Company and Burnham and Morrill. The responses ranged across the spectrum. Donald Farquharson, a long-time Liberal Island politician and packer, argued that Canada should shut out "rich foreign firms of Americans." At the other end of the gamut, Cape Breton politician Dr. Arthur Kendall noted the importance of large American firms and expressed fear that any interruption of these firms' operations might lead to potential trade retaliation and shut the American market to Canadian lobster exports. Moreover, Kendall understood the strong tentacles of big companies' power (and emerging practices in leasing facilities to locals) in concluding that the "American packers named could easily operate canneries in the names of Canadian packers."[49]

If the roles of politicians and packers often overlapped, opponents were always on the lookout for preferential treatment. In August 1894 Ephraim A. Bell of Cape Traverse, Prince Edward Island, wrote to the Liberal *Daily Patriot* with a not so veiled reference to what was widely perceived as favouritism towards Acadian packers linked to the Conservative Party. He demanded that lobster regulations be enforced "regardless of creeds, nationalities or political distinctions." He claimed that the Conservative government, in an election year, was not putting a stop to illegal fishing "among its friends ... they want to get votes and lobster will get votes just as well as" large infrastructure projects such as local bridges, locks, and dams in Quebec and Ontario. Not content with gerrymandering Prince County, in a lovely turn of phrase he accused the government of "gerrymandering the lobsters around its shore."[50]

Political interference could also target the policing of specific packers. A report produced by the St. Francis Xavier University Extension Department in the 1950s claimed that although difficult to substantiate, a certain packer who "was rather free with 'hot' money" derived from lobster poaching had used it to buy "votes in support of the incumbent's opponent." His efforts were not successful and after re-election, the Member of Parliament "declared war on the packer with the result that protection officers applied the pressure and their efforts were effective."[51]

The link between federal elections and the extension of lobster seasons was perhaps the most blatant use of lobster regulations for political ends. In a 1921 discussion of illegal lobster packing, Inspector S.T. Gallant of Prince Edward Island concluded that "election years saw the worst abuses."[52] Fifteen years later, Liberal politician and packer Magee noted that when an election was upcoming, a seized boat might be returned, and no conviction carried out.[53] Eleven days before Maritimers went to the polls in the tightly contested 1896 federal election, the Conservative minister responsible for fisheries extended the lobster season. The season was to end on either 1 or 15 July, but a circular issued on 12 June extended the season for fifteen days "in view of the backward season and in consideration of the strong grounds urged by all those interested in the lobster season." The motivation, however, seems to have been an unsuccessful attempt to prevent the defeat of the government.[54]

Enforcement of the regulations was also directly connected to electoral prospects. An 1894 letter from Conservative James H. Whitman of Salmon River, Halifax County, explained that local officials had never enforced the close season until the fall of 1893. That year, the Fisheries Protective Service cutter visited the area and smashed illegal traps. The result in Salmon River was that twenty fishermen who had always voted Conservative, voted Liberal in the March 1894 Nova Scotia provincial election. Destroyed traps changed political allegiance. The canning sector, at the time, was almost exclusively associated with the Liberals who, according to Whitman, "preach that the tories are the party who made the law and will not allow them to catch lobster to help feed their families." Gloucester Conservative Member of Parliament Théotime Blanchard raised the same alarm. He wrote Charles Hibbert Tupper, the unusually diligent Minister who tried to uphold the regulations, bluntly declaring that the enforcement of lobster regulations was "destroying the chances of the Conservative Candidates at the next general election."[55]

Notwithstanding many examples of politicians influencing enforcement to their own advantage, it is important to note that they were not always successful. In the Spring of 1904, W.S. Fielding wrote to the Minister responsible for fisheries in support of a lobster season extension, adding that if there was "any disagreement" he "would like the opportunity to discuss it," thereby keeping the door open for further political leverage. Fisheries Commissioner Edward E. Prince was consulted but did not see any justification for the extension, although this request occurred immediately before a general election.[56] A similar strategy was employed at the same time by Robert E. Armstrong, the defeated 1900 Charlotte County Liberal candidate, who wrote to the Minister of Marine and Fisheries. With the prospect of many Grand Manan Islanders having to move to the United States after a miserable season in 1904, Armstrong insisted that relaxing the local minimum size limitation would aid local Liberal support.[57] The existing policy held.

The sensitivity around elections – as moments when lobster seasons might be extended or prosecutions delayed or dropped – speaks to the political importance of the lobster fisherman's vote. The influence was also reflected in the widespread use of petitions as a tool for shaping policy. Both fishermen and packers used petitions to signal popular support or opposition to specific regulations, with the number of signatures interpreted as a measure of electoral favour. In practice, packers often wrote, printed, and distributed petitions to fishermen under the assumption that sheer volume mattered. But this is questionable. In fact, not every fisherman signed with complete autonomy, as many were beholden to specific canners – either through debt or dependence on them to purchase their catch.

There was a constant stream of petitions for altering the dates of the open season and in favour of fishing season extensions with peak waves such as the aftermath of the new 1899 regulations.[58] Community-based petitions signed by fishermen – sometimes with an X and sometimes in identical cursive to the signature above – had standard text and were usually typeset. These identical templates suggested the initiative of either packers or exporters. In 1894 the Minister of Marine and Fisheries received a petition from the lobster fishermen in Framboise, Cape Breton, requesting a season extension in the distinct and recognizable handwriting of their main local packer H.E. Baker.[59] In March 1910, after Commander Wakeham's recommendations to shorten the open season and remove the size limit became public, a standard petition was distributed by packers to their fishermen in Antigonish, Guysborough, Inverness, Pictou, Lunenburg, and Queens in Nova Scotia and in Prince County, Prince Edward Island. Some communities returned pages of names, while a petition from Tancook Island in Lunenburg County included a single signature. The petition was directed at fishermen who owned their own boat, gear, and traps and who claimed their income would suffer with a curtailed season and no legal option to sell small lobsters locally to packers.[60] A letter from a fisherman from Western Head in Queens County published in the *Halifax Herald* at this time stated that the Liberal-connected packers were circulating the petition to remove the size limit and encouraging fishermen to sign it. The author – evidently a Conservative – held that the sale of small lobsters went against the interest of lobster fishermen and concluded with a rousing plea that "the fishermen's sword is his vote, Queens-Shelburne is one of the counties where it can be used with good effect."[61]

The coordination among canners in circulating petitions should not be surprising as given they were colluding to fix the prices offered to lobstermen.[62] In addition to these private cartels, packers also organized in public formal associations such as the Canned Fish Section of the Canadian Manufacturers' Association. This organization, together with the Canadian Fisheries Association was especially visible around the First World War when it began publishing a monthly journal, *Canadian Fisherman*, to publicize its perspective. The Canned Fish Section took constant aim at politics and politicians as the root of the crisis in the lobster fishery. In 1918, the Section adopted a position that lobster regulations be removed from "the realm of politics" by enshrining lobster regulations within statutory legislation rather than have them governed by the caprices of Order in Council, where extensions to a season could be granted easily to coincide with any general or by-election.[63] The Canned Fish Section of the Canadian Manufacturers' Association, in its testimony before the Maclean Commission, identified the main problem of the lobster fishery not as illegal or overfishing, but "political interference." It held both Liberals and Conservatives responsible for partisan activities and identified politicians as "catching votes" during elections by exploiting differences between packers and fishermen. In hyperbolic prose, section president R.H. Williams declared, "The Politician with his interference is more responsible for the decline of the lobster supply than either fisherman or canner."[64]

Williams's language became even more extreme when representing the interests of packing wholesaler Roberts, Simpson and Company. In a 1927 speech to the Halifax Rotary Club, he criticized the 1873 *Fisheries Act* as designating enforcement to "a cheap class of ignorant partisans, sharing local prejudices and wielding their power in accordance with their own petty local preferences."[65] Lobster fishermen might make similar claims of the industry.

Packers also acted at the individual level. Michael Neville, of Halifax, who operated canneries on Nova Scotia's South Shore, regularly wrote the Minister of Marine and Fisheries to flag his own concerns. When groups of fifteen individual fishermen were given the opportunity to gain licences for cooperative canning in 1907, Neville wrote Minister Louis-Philippe Brodeur, explaining that existing canneries were already losing between $2,000 and $3,000 every season and that the Canadian industry would be completely ruined as had been the case in the United States. According to Neville, existing canners had the right to have their investments "safeguarded" and "fishermen cannot be fishermen and fish curers."[66] Other packers were also in very regular contact with politicians and department officials in Ottawa. W.F. Tidmarsh, manager of Island operations for the Portland Packing Company, regularly offered unsolicited advice. In 1909 he took the initiative to send the Minister of Marine and Fisheries a copy of the resolution debated by the Prince Edward Island legislature regarding the most appropriate local open season, a matter well outside this house's constitutional jurisdiction.[67]

The organization of packers and canners was mirrored by similar associations composed of fishermen. When Maritime inshore fishermen came together in common cause, it was usually connected to lobster. Yet even when organizations appeared to be a grassroots movement, the state was at least tangentially involved by supporting these groups with special legislation or permits. At the turn of the twentieth century, two concurrent and often similar fishermen's organizations developed at the community level – cooperatives and the Fishermen's Union of Nova Scotia. Both associations were a direct response to the trust actions of lobster canners.

At least one example of community-based canning operations predated this wave. In Lower East Pubnico, Nova Scotia, a factory led by merchant Henry T. d'Entremont was constructed in 1886. This cannery brought together twenty-five local investors composed of d'Entremont and local fishermen, and the profits were divided on the traditional share system. The division was exactly the "co-adventurer" model used for shipbuilding, ship ownership, and most extended fishing trips. For this cannery, the weight of individual catches determined the proportion of the share.[68]

After a moratorium on new cannery licences was introduced in 1898, the Dominion government permitted exemptions for community groups wanting to pack lobster. In 1904 local members of Parliament, such as Prince Edward Island's Donald A. McKinnon, passed along letters from fishermen who wanted to pack "in order to protect themselves from underpayment for their lobsters by the canners."[69] In practice, very few groups adopted this strategy. By 1908 only two of the 127 canning licences in New Brunswick were held by cooperatives, while eleven of the 183 licences in Prince Edward Island fell into this category.[70] In an attempt to promote this model, the Minister of Marine and Fisheries announced in 1909 that if no fewer than fifteen lobster fishermen formed themselves into a "co-operative association to can their own lobsters caught by them and agreed to share alike in profits or losses" they would be granted a licence subject to cancellation if the conditions were not followed.[71]

In the first decade of the twentieth century, fishermen's associations traced their roots to a hybrid form of organized labour and a business lobby association. They reflected both contemporary critic Colin McKay's description of the Fishermen's Union of Nova Scotia as being composed of "small capitalists as well as workers" coming together to protect their vulnerable resource and fight against the "New England fish trust" and historian Brian Payne's understanding of fishermen as both capitalist producers and resource stewards.[72] In the spring of 1905, there were rumours that the socialist American Labor Union had begun organizing lobstermen in Maine and the Maritimes. There were probably fewer than 3,000 lobster fishermen in Maine compared with the over 30,000 lobstermen in the Maritimes, making the region essential for any control of the commodity.[73]

The rival American Federation of Labor began its own at least temporarily successful unionization campaign in Maine and sent one of its Canadian organizers on an unsuccessful six-week trip to Nova Scotia.[74] The organizer's visit coincided with the establishment of the Fishermen's Union of Nova Scotia, which historian Robert Babcock concluded was "designed to keep the farmers of the sea out of the regular trade-union movement."[75] It united inshore fishermen, primarily of lobster, who owned their own boats and gear and perceived themselves as independent entrepreneurs. At its 1906 annual convention, it formally rejected overtures from trade union organizers, but sociologist Gene Barrett may have misunderstood the organization in describing it as simply "merchant-controlled."[76] Some buyers and shippers were involved, but local chapters were strikingly non-partisan, willing to both cooperate with and criticize the Liberal governments in both Ottawa and Halifax.

The individual behind the Fishermen's Union of Nova Scotia (FUNS) was prominent Liberal Moses Nickerson. He was a onetime employee of the Portland Packing Company, former newspaper editor, a past member of the 1898 Royal Commission, an early broker for shipping live lobsters to Boston, and a Liberal MLA. Under a 1905 Nova Scotia act he sponsored a bill to permit groups of fifteen fishermen who could legally incorporate as "stations" for the purpose of owning tax-free common property, using their collective buying power to purchase gear, enhance education around packing lobsters for the export trade, and, in some communities, cooperatively pack lobster. In 1909 there were, according to Nickerson, forty stations with memberships ranging between fifty and 150 men.[77] No progress or growth was made after 1909, and by 1913 the organization had disappeared without a trace in the records.[78]

Events in Gabarus, Cape Breton, in May 1909 put the FUNS front and centre. The local station led a strike demanding the price of $3.25 per hundred lobsters instead of the $2.35 offered by the buyers. The Gabarus men were joined by fishermen at nearby Main-à-Dieu and communities in Antigonish County and the Gulf Shore of Cape Breton, totalling between 250 and 300 men altogether.[79] The immediate sacrifice to the lobstermen was high, as the open season was brief and live lobsters could not be stored. The intermediate cost was even higher, as Burnham and Morrill mothballed its local factory for the 1909 season, leaving no buyers in the vicinity. But even before the strike had begun, the Gabarus station had inquired into getting its own canning licence, and a cooperative cannery was established there and at Gull Cove.[80] Overall, the FUNS may have rejected the model of organized labour, but local stations did not. The Fishermen's Union of nearby Long Point and Judique requested that the Department of Marine and Fisheries appoint a Board of Arbitration to settle the price canners should give, suggesting a model along the lines of labour arbitration.[81]

Local chapters of the FUNS in western Nova Scotia, oriented towards the live export market in Boston and not locked in by monopoly buyers, were more interested in resource stewardship. Some stations petitioned unsuccessfully to expand a scheme piloted in Cape Breton in which seed lobsters were purchased from fishermen at a price above the market rate and stored for the season in special pounds in the hope that upon their release they would reproduce.[82] In December 1905 individual stations wrote the Minister of Marine and Fisheries expressing their concerns about packers, violations of the law, and the impotence of fishery officers. They did not use a boilerplate template letter but addressed specific local matters. The Station secretary from Little Harbour, Shelburne County, complained that "the illegal destruction of seed and undersized lobsters is being carried on by canneries and fishermen on this coast to an alarming extent," with fishery officers "doing comparatively nothing to stop illegal fishing and canning." The letter indicated that the station had written to Andrew Robertson, the local Inspector, requesting a meeting so that "we may assist him in providing ways and means for strictly enforcing the present law and it seems impossible for us to get his assistance." Therefore, the letter concluded with a request that the Minister "use his influence for the protection of the lobster fisheries and if necessary, compel his officers to strictly enforce the law."[83] Similarly, forty fishermen from Station Number 5 at Port Mouton, Queens County, expressed concern about a local factory packing undersized lobsters. Fines, according to the fishermen, should escalate with each conviction, and after a third offence "nail up his factory." Clearly expressing their personal stake in the health of the local resource and their desire to be better informed about what protection was available for lobsters, the letter concluded, "it seams that our fisheries officers cant stop it will you send us the act of the laws of the fisheries as have no guide to go by."[84] In February 1912 the Little Harbour, Shelburne County, station again wrote to the Department of Marine and Fisheries and expressed concerned of lobster's "eminent danger of being exterminated." It asked that the laws around the close season be enforced, that no lobsters under ten inches be caught, and that something be done about the Americans anchoring three miles off the coast and buying lobsters during the close season.[85] For lobster fishermen engaged in the live export market, as both small capitalists and workers, advocating for conservation goals against packers was a communal political act.

Cooperatives, on the other hand, appeared to be oriented towards creating an alternative economic and social structure that ultimately freed lobster fishermen from the monopoly power of packers. Here we see lobster fishermen clearly as both workers and small entrepreneurs using their collective resources to compete with larger corporations. Before the 1909 Marine and Fisheries Committee respecting the lobster industry, R.N. Venning, Superintendent of Fisheries, testified that outside of Nova Scotia and with its provisions under the FUNS, cooperatives such as that in Launching, Prince Edward Island, were not incorporated. These cooperatives packed lobster under a special licence but remained closely tied to corporate canners who maintained their monopoly on specialized supplies and distribution networks.[86]

After the First World War a scattering of new cooperatives slowly emerged, each rooted in specific local circumstances – much like the earlier Fishermen's Union. Fishermen at Tignish, Prince Edward Island, were frustrated with the low prices they received for their lobsters. The lack of competition on Prince Edward Island meant that the prices were a third of what fishermen on the mainland earned. In response, Tignish lobstermen in 1920 met to find solutions. Like most fishermen, those in Tignish had no access to the credit necessary to build a factory or buy supplies. The men, therefore, approached local lawyer Chester F. McCarthy, who they knew to be sympathetic. Originally, McCarthy agreed to finance the entire construction of a cannery in exchange for a percentage of sales. Later, the cannery was reorganized, with McCarthy as general manager and thirty-six fishermen forming a joint stock company (incorporated in 1925) with the purpose of both canning and buying fishing supplies. Membership was restricted to those who owned their own boats and gear and could raise $200 for a share. In January 1926 the first dividend of 7 per cent was distributed.[87] By 1927 there were thirty or thirty-five shareholders at the cooperative – sometimes described as a union – at Tignish busy canning their own small lobsters and selling live lobsters through Paturel in Shediac.[88] Their canned lobsters were forwarded to the Prince Edward Island branch of the Portland Packing Company and February 1931 minutes indicated that the cooperative was in contact with the Prince Edward Island branch manager for advice on the quality of the pack.[89]

About the same time in 1923 as the Tignish cooperative was established, Father J.J. Tompkins used his own resources to help establish a cooperative community cannery at East Dover, Guysborough County. This operation would be the precursor to the string of canneries established in the 1930s by the Antigonish movement.[90] The 1927 Royal Commission Investigating the Fisheries of the Maritime Provinces and the Magdalen Islands recommended the establishment of cooperatives as the most important strategy to improve fishermen's standard of living. Reverend Dr. Moses Coady of St. Francis Xavier University's newly established Extension Department was seconded in 1929–30 by the Department of Marine and Fisheries to organize fishermen. After 1930 the newly formed Dominion Department of Fisheries (along with the Carnegie Foundation) was one of the primary funders of this influential social movement. When local chapters of organized fishermen and cooperatives met in June 1930 to federate into the United Maritime Fishermen, Tignish's Chester McCarthy was the initial president.[91]

Corporate canners and local merchants were highly critical of this initiative and directed criticism at the competency of "fishermen canners."

In 1929 the industry journal *Canadian Fisherman* shifted blame from the small independent canneries to cooperatives that ruined the reputation of the industry and resulted in "less earning for everyone!" Rather than cooperating among themselves, "Fishermen need to cooperate with the packers."[92]

Cooperatives were also involved in the lucrative live export market, transferring money from traditional buyers to the fishermen themselves. In 1937 lobsterman William Feltmate of Whitehead, Guysborough County, received a cheque for $32 from the cooperative, $22.20 more than what he would have earned had he sold his live lobsters to the local merchant. By the mid-1930s packers retaliated so that they refused to buy the small lobsters used for canning from individual fishermen who sold their larger lobsters to cooperatives for better prices.[93]

It is hardly surprising that cooperatives proved popular in many communities. By 1936 *Maclean's* magazine reported that there were seventeen lobster cooperative canneries serving seventy communities, primarily in eastern Nova Scotia areas with Acadian, Irish, and Scottish Catholic populations. In the community of Havre Boucher, the thirty-five fishermen earned $7,000 more from their catch by selling to the local cooperative rather than had they sold to an independent packer.[94] Cooperatives in eastern Nova Scotia, working together with St. Francis Xavier University scientists and the Nova Scotia Department of Marketing, were able to solve the issue of high mortality that had made shipping live lobsters to Boston from eastern Nova Scotia difficult. Now the valuable large lobsters could be exported live, while the smaller "canners" packed locally, leaving more money in the local community. Other strategies for economic justice were also pursued, and in May 1930 it was reported that Glace Bay lobstermen threatened to strike.[95]

Reverend Coady drew attention to "the bond" between St. Francis Xavier University's Extension Department and both the Provincial and Federal governments. Employees of the Extension Department were required to renounce partisan politics, and its leaders, such as Moses Coady, kept close watch on internal political machinations, including plots to remove sympathetic fishery ministers from their office.[96] While the Antigonish movement attempted to be non-partisan, it was staunchly sectarian. Predominantly Catholic, the movement included only a few Protestant individuals, hired strategically to build bridges to the Protestant community.[97] This sectarian difference was reinforced by geography, for apart from Nova Scotia's French Shore on the Bay of Fundy and urban centres, the Maritimes' Catholic populations mapped closely to the areas where the canning sector remained the most important. The Antigonish movement and its cooperatives were led by Catholic priests out of a Catholic university, and local community operations were financed by local priests or parish wardens. The astute Coady wrote to the Minister of Fisheries in 1936 emphasizing the need to hire non-Catholic organizers "to mitigate the feeling of the movement being a Catholic movement," but there was limited success in Protestant communities.[98]

While the Antigonish movement was based in eastern Nova Scotia, the United Maritime Fishermen's cooperatives expanded in New Brunswick's Catholic Acadian communities. By 1942 the United Maritime Fishermen operated sixty-seven canneries, about thirty of them in New Brunswick's French-speaking villages.[99] On the Gulf of St. Lawrence, Acadians formed the highest proportion of fishing communities, and ethnic politics with the English-language Antigonish movement proved complicated in a new wave of Acadian nationalism. The Catholic cooperative movement split along language lines in the 1940s.[100]

By the 1930s, lobster had become the cornerstone of the United Maritime Fishermen's cooperative movement, prompting a shift from education to marketing.[101] This reorientation reflected the growing importance of the live market trade. Advances in communication and transportation technologies expanded the geographic reach of the trade, and by 1938 cooperatives were established off the west coast of Newfoundland to ship live lobsters to Gloucester, Massachusetts.[102]

Following the Second World War, the Prince Edward Island Fishermen's Association formed in 1952 as another local organization, embedded in its distinct culture and community concerns. This English-language inshore fishermen's organization was a "Voice of the Fishermen" offering to cooperate with the provincial Fisheries Development Committee to form a joint plan to provide disaster assistance without political patronage or direct relief.[103] In 1953 it formally responded to proposed lobster regulation changes that would limit the number of traps individual fishermen could fish. Its early meetings focused almost exclusively on lobster regulations, enforcement, and conservation and involved both the federal lobster fishery scientist Dr. D.G. Wilder of the St. Andrews Atlantic Biological Station and Catholic clergy from Charlottetown's St. Dunstan's University.[104] The business side of the lobster industry celebrated its focus on illegal fishing and the related depressed prices the black market generated.[105] The state and the church remained entwined even as this organization was both a lobby and a forum for consultation on government policy.

The involvement of either the provincial or federal governments in various local fishermen's organizations in providing a legal framework, offering financial support, or providing expertise muddies the notion of them being purely grassroots fishermen's organizations. This reflects the way cooperatives were cast in the 1930s as the "middle way" between laissez-faire free markets and state socialism, and many inshore lobster fishermen self-identified as both workers and small capitalist producers.

Conclusion

Lobster politics in the Maritimes moved from being close to the centre of electoral and partisan politics to becoming increasingly linked to associations, albeit with connections to the state. Until after the First World War, lobster fishermen's votes mattered a great deal, and a significant number of regional politicians were personally involved in the industry. As a result, regulations were connected to electoral success, as they literally "caught votes" in closely contested federal elections among fishermen and their households. But the importance of the lobster question to federal politics declined over time. The political weight of the region diminished as seats were redistributed and the size of the House of Commons grew. Just like there was a "lived law" there was a "lived politics" where local politicians maintained and distributed power and privileges to their "friends," and the importance of this power was significant with a relatively weak and distant state. Local politicians dispersed patronage, interfered in the judicial process, and lobbied on behalf of political friends trying not to alienate voting fishermen. Part of the political strategy used by both Liberals and Conservatives was exploiting and capitalizing on differences between packers (and later buyers) and fishermen as both packers and lobster fishermen played politics for their own economic, social, and environmental goals. Packers and lobster fishermen also pursued their own agenda through their own associations.

During the 1920s and 1930s fishermen used the state to fund cooperative outreach and education. By the 1950s the federal state was using fishermen's associations as consultation forums to claim legitimacy on fisheries policy decisions. As the impact of fishermen's votes in many ridings declined, lobster politics changed. It still might be important to "catch votes" in certain ridings, but lobster was no longer the centre of Atlantic politics, and lobster politics became framed within other regional political concerns, such as development and a decent standard of living.

Chapter Seven

Lobster Regulations as Social Policy, 1929–1970

Social and economic policies were always embedded in lobster regulations; the laws were never solely about conservation. Prior to the Second
World War, however, concerns about the overall health of the lobster
stock tended to dominate both department and political discussions.
The "lobster question," which took up so much political space, was as
much about lobsters as it was about the canners, fishers, factory workers, and communities connected to the fishery. In the postwar years,
this focus shifted. Lobsters themselves increasingly took a back seat
to people-centred concerns, framed within broader agendas of social
and economic development. By 1955, for example, the *Canada Year Book*
justified the enforcement of lobster regulations not by referencing ecological sustainability, but by warning that "economic depletion could
result" from laxity in policing.[1] The framing of the problem in terms of
economic consequences rather than stock vulnerability was significant.

The modernization impulses of the stronger federal state were
increasingly focused on reconciling the tensions between economic
efficiency and social welfare. In this context, the "lobster question"
largely faded, replaced by a broader policy emphasis on increasing
fishermen's incomes through labour efficiency, technological advance-
ment, and the privatization of access rights. For a time, lobster fishing
itself nearly disappeared from federal policy discussions, as departmental attention shifted to the capital-intensive, industrialized offshore
fin-fish sector. Yet, in seeking to reconcile the contradictions between
modern resource extraction and a "social fishery" economic thinking
of the day recast lobster from an open access commons to a form of
state-regulated private property. In the late 1960s lobster became the
first Atlantic fishery to restrict entry through a licensing system, and
by 1976 all East Coast ocean commercial fisheries followed suit. The
closure of the commons had especially significant consequences for

lobster, it was the most valuable inshore Atlantic fishery, the primary source of fishermen's earnings, and the easiest fishery to enter prior to restricted licensing.

When the fishery was open access, the value and widespread presence of the lobster fishery meant that conservation regulations were regularly altered or ignored in response to economic crises or short-term material needs of local communities. Before the Second World War, lobster regulations were part of the very limited toolkit of a weak, but "sympathetic state" and the short-term well-being of communities and packing interests almost always trumped the welfare of lobsters. In the postwar period, the expansion of income support such as Unemployment Insurance, indemnity programs for gear, fishermen's boat loans, and eventually buyout licence packages allowed the federal state and its local agents to separate community welfare and sustainability from Department of Fisheries' concerns about resource sustainability. The most important fishery welfare programs generally were transferred to other departments. There is a small literature that examines the fishery and welfare policy. Fred Winsor explored how Nova Scotia's offshore fishermen were added and removed from the Provincial Workmen's Compensation Act in the 1920s.[2] Others such as economist William Schrank have examined the impact of the introduction of fishermen's benefits to Unemployment Insurance in the 1950s but, generally, resource management has been an overlooked aspect of social policy and welfare concerns.[3]

Although the protection of an important economic natural resource was at the centre of lobster regulations, state management of the fishery always had social policy goals. Prior to the Second World War, for example, emergency financial relief or efforts to halt the stream of outmigration to the United States influenced policy decisions. These two concerns were of course enmeshed, as the most obvious solution to address local poverty was leaving for better prospects elsewhere. Population growth was an important measure of the success of a society in the nineteenth century, so keeping young men from leaving for the "Boston States" carried social and political weight. Digby fish merchants wrote to Minister of Marine and Fisheries Charles Hibbert Tupper in 1894 and argued that a six-week extension for the lobster fishery and the abolition of size requirements would be enough to create the conditions for young men to stay at home.[4]

Michele Landis Dauber, in her 2013 book *The Sympathetic State: Disaster Relief and the Origins of the American Welfare State*, argues that the tradition of American government intervention at times of natural disasters, be they hurricanes, earthquakes, floods, or fire, formed a precedent for those constructing the welfare state, as these calamities inflicted great harm on individuals through "no fault of their own." The notion of not being responsible for bad fortune was connected to the very foundations of modern welfare in which people were divided between the deserving and undeserving poor. The deserving poor were worthy of charity and support as their poverty was not the result of bad character or habits.[5] The same moral-based logic was applied to altering lobster fishing regulations in response to immediate need. Yes, political intervention might be motivated by electoral success as discussed in the previous chapter, but it could also find its roots in humanitarian concern. In a pre-welfare state era, the lobster fishery functioned as part of the state's informal toolkit for delivering social assistance and income support. Fishermen suffering poor catches were characterized as hard-working, independent, white men striving to feed their families and pay their bills. In such extenuating circumstances why should an arbitrary law that dictated when people might legally catch and sell lobsters or deemed certain lobsters too small to keep, carry weight in either a moral economy or in public consideration? Inshore lobster fishermen, framed as deserving assistance, lived either under the limited provincial poor law regimes of the Maritimes or relied on private charitable support – religious or ethnic – in Quebec. Both systems placed the burden of care on already strained small local communities. In this context, pressure on other levels of government to offer relief required flexible and improvised solutions before the expansion of post-Second World War income support programs. The federal Department of Marine and Fisheries and later the Department of Fisheries was not the only department to adopt ad hoc solutions. Farmers were the recipients of compensation and Parks Canada historian Meg Stanley has noted that during the First World War Banff National Park increased the number of timber-cutting permits, grazing permits, and fishing licences to assist local residents in dealing with the dramatically inflated costs of fuel and food.[6]

Federal and provincial politicians and bureaucrats were regularly placed in a situation where the well-being of the local community was at odds with the preservation and sustainability of lobster stock. As noted earlier, ecological factors such as ice conditions, the hardness of shells, and breeding cycles played a role in the establishment of close seasons, but the local human economies were also considered. Politicians from Nova Scotia's southwest coast promoted the winter lobster fishery as it coincided with the period when "a very large number of our fishermen are idle." Similarly, a fall season was introduced in the Northumberland Strait in part to deal with late spring ice but also so as not to interfere with spring planting for the farmer/fisherman.[7] The overall importance of the lobster fishery to coastal communities was clearly understood. In the 1920s the *Montreal Standard* explained to its readers that for fishermen in "many districts the earnings from the lobster catch in many cases represent the margin of difference between a tolerable standard of living and a condition which few would care to put up with."[8] Without a healthy lobster fishery, outmigration and poverty would have been even more severe.

So far, lobster conservation regulations have been cast as a contentious and partisan political issue. As the dates of lobster seasons were not set in statutory law but were Orders in Council, it was simple and efficient to assist local lobster fishing communities in need with an extension of the open season. Archival files from the Department of Marine and Fisheries contain numerous petitions and requests from fishermen, canners, clergy, and merchants, all citing "exceptional" circumstances. Almost every appeal invoked the spectre of outmigration and privation faced by those who remained. It is also one of the places that one can catch glimpses of individual lobster fishermen in their pleas to the Department. A petitioner who signed himself anonymously "Fisherman" from East Dover, Nova Scotia, warned against the proposed 1910 regulation changes. He urged department officials to "travel around our shores in winter and see the familys that are kept in bread and clothes from the Lobster fishery" and cautioned that new regulations would lead to the government having to "feed the fishermen."[9]

The words of fishermen were not all that different from those of politicians and clergy. A cold and wet late spring in 1897 left Acadian farmer-fishermen in the Westmorland area with their potato crops rotting in the ground and insufficient time for a second planting. This led local politicians to plea that misery could be prevented the following winter with an extended lobster season. Liberal New Brunswick MLA (and future premier) Clifford Robinson wrote directly to the Commissioner of Fisheries in 1897 after the request for an extension had been declined. He noted that his district had about ninety factories, three hundred boats, and employed between 1,200 and 1,500 men (apparently the women factory workers were not relevant). The lobster fishing season was a disaster, and the men had not been able to clear their debts from the previous year. The predicted result was that the area would be "in a deplorable state next winter."[10] When both the mackerel and cod fishery failed in the Magdalen Islands in 1903, a local priest wrote to the Minister of Marine and Fisheries warning that there would be a famine if locals were not permitted to fish lobsters out of season in the fall.[11]

The poor and isolated population of the Magdalen Islands may have been especially vulnerable to the threat of famine. In September 1905 a telegram was sent to the Minister of Marine and Fisheries claiming that fishing had been a total failure that summer and that the people were "threatened with much privations and hardships during the coming winter."[12] The fishermen requested an extra open season for lobsters and that they be allowed to fish them in the "lagoons," which were supposedly a permanently closed protected breeding area. A fall lobster fishery outside the lagoons was rough and dangerous, and the prospect of a difficult winter and immediate safety of the fishermen took precedence over conservation. The "privilege" of an extra season was granted in the Magdalens in 1904 and 1905, which led to fishermen, canners, and merchants in all other areas to make parallel requests based on similarly "good grounds."[13]

Economic hardship and a widespread understanding of a moral economy following the First World War meant that the Department of Marine and Fisheries regularly placed people above fish. In the early 1920s petitions from Protestant and Catholic religious leaders from "destitute" communities west of Halifax led to the introduction of an additional six-week fall fishery in 1921.[14] The increased visibility of the poverty of inshore fishermen played a central role in the creation of the 1927 Maclean Commission, where the connection between community and household poverty and the lobster fishery was a recurring theme in the testimony. For example, a parish priest in Arichat, Cape Breton, testified to the poverty of the lobster fishermen stating that their living conditions were "not much better than the standard of living of the Eskimo and the Indians." He noted that his parishioners rarely had fresh meat, had no milk (even in their tea), and that the local children were visibly undernourished and did not attend school because of inadequate winter clothes. In this description of poverty, he cast the fishermen as deserving better as evidenced in their work habits. These men were not lazy – they began their days at 2:00 a.m. and risked their lives to provide for their families.[15] It is difficult not to be struck by the extent to which this example was steeped in the rhetoric of those deserving charity – their race, work ethic, status as fathers, and the presence of vulnerable children justified assistance.

The Commission in its support for cooperatives rejected the earlier, often inferred welfare claims of merchants and packers who had sometimes cast themselves as part of the charity network. Before the First World War, a prominent Cape Breton lobster packer and merchant, H.E. Baker, approached the Minister requesting an extension to the lobster season, stating that without the opportunity to fish longer local merchants would "become compelled" to halt credit, leaving "a large number almost entirely destitute for the coming fall and winter."[16] The assumption that merchants should not bear the risk of tough times but only profit from good years is striking.

Fishermen also were explicit in their requests. When lobster fishermen in Wedgeport, Nova Scotia, asked for a special fall season, they did so as "relief for the winter of 1931–32."[17] While the special season was granted in 1931, it was not repeated the following year, leaving those on the South Shore angry, along with a group of Cape Breton fishermen who had requested the same with the expectation of an exemption being granted in the name of relief.[18] Many lobster fishermen responded by fishing illegally. From the perspective of 1936, a newspaper reported that, "The politicians being human and not having the will to attempt a constructive attack upon the depression, conceded the demand some years."[19]

When lobster fishermen were caught and found guilty of fishing out of season, there was not always the community or political will to prosecute or collect fines. Requests for special consideration may have had partisan pull, but the power of a moral economy was also evident. Petitions requested that fines be waived for certain canners and fishermen convicted for breaking the law. In October 1895 this was the case for a Prince Edward Island packer described as "a very poor man with a large family" who "scarcely got any lobster" during the open season so continued to fish and canned after it had ended. His supporter argued that he was "anxious to get enough to pay his bills and have something for his children to eat." One Prince Edward Island federal fishery inspector complained in 1922 that "even clergymen intervened on behalf of lawbreakers who pleaded poverty as an excuse." Members of Parliament also sympathized and held that, considering their "hardship and disappointment," fishermen were not "altogether to blame" for breaking the law.[20]

Reestablished by the Liberals in 1930 to regain waning regional support, the "new" Department of Fisheries alongside its conservation objectives took a central role in addressing the living standards of lobster fishermen and struggling coastal communities. Under the succeeding Conservative government, the federal government operated a subsidized boat to collect live lobsters in season along Nova Scotia's Eastern Shore from 1931 to 1935. The goal was to improve incomes by giving fishermen direct access to the more lucrative American live lobster market. In this geographically isolated region, with no railways and poor roads canning lobsters remained the primary local option.[21] More broadly, during the 1930s, the construction of breakwaters and wharves also served as a means of delivering patronage-based relief through the Department of Public Works. The Royal Commission on Maritime Claims, established in 1926, recommended federal assistance for constructing small-boat harbours, wharves, and breakwaters as "a great benefit to our people" in the midst of "hardships."[22] So, for example, in March 1934, while advocating for the construction of a break-water in his South Shore Nova Scotia constituency, opposition member Liberal James Ralston integrated arguments about moral worthiness for assistance into a public works program. He noted that employment was greatly needed in this location and "It was a community of people which have tried to work out their salvation."[23] The decision to build in a particular place was here linked to the moral worth or deservingness of the community.

During the 1930s, politicians in Ottawa also wrestled with how best to support lobster fishermen when storms destroyed their boats, traps, and gear. In the spring of 1934, Ralston advocated for the creation of some "extraordinary emergency provision," noting that many lobster fishermen had lost the very gear they needed to make a living. He reasoned that, while such aid could technically fall under the Relief Act, the Department of Fisheries should supply the means by which men could re-equip themselves. "They are not asking for the dole," he insisted, "but for an opportunity to help themselves." Ralston renewed his appeal in February 1935, after another winter storm caused simi-lar losses. He maintained that these fishermen were not only with-out equipment but also lacked access to credit, and that government assistance could avert widespread unemployment and dependency on relief. This time he reached back to 1882 for a precedent: a provincial grant made by Nova Scotia that year to help fishermen replace lost gear after a storm.[24]

When the federal Liberals regained power in October 1935, Ralston's argument had a more sympathetic hearing from the government. However, not all within the administration agreed. Minister of Fisher-ies Joseph-Enoil Michaud opposed using departmental appropriations as welfare payments something he believed fell under provincial and municipal responsibility. He considered the $300,000 appropriation made by his department inappropriate, stating it "could have been placed just as well under the jurisdiction of any other department. I submit that we cannot assume jurisdiction of those individuals who give only part of their time to fishing." At this point Michaud estimated that 20 per cent of the population of the Maritimes was dependent on fishing for its livelihood but maintained that this did not make these people the department's responsibility.[25]

A single devastating storm in early December 1938 prompted another $100,000 to be appropriated from Fisheries to compensate for lost boats, traps, and gear. Cost-sharing payments were arranged for lobster fishermen in New Brunswick and Prince Edward Island, but delays arose for those in Nova Scotia, as the province refused to support any indemnity program. Much like the welfare programs historian E.R. Forbes examined during the same period, this form of fiscal federalism required provincial spending to unlock federal funds.[26] Nova Scotia, however, went further, pressing for fisheries funds to be used for the cost-sharing of road construction in storm-affected districts.[27] Meanwhile, local politicians, such as Liberal Vincent-Joseph Pottier – the first Nova Scotian Acadian to be elected to the House of Commons – voiced the urgent need for trap restoration. Pottier estimated that 44,440 lobster traps had been destroyed in his riding of Shelburne, Yarmouth, and Clare alone. In his support for government loans to replace destroyed goods, he echoed language used to justify public relief to the deserving poor:

> Our fishermen have resourcefulness, courage and pride that have enabled them in the past to keep off relief. One family would help another; one relative would help another, and the young would support the old. ... They have believed in being able to manage and control their own livelihood, rather than in depending on welfare workers. They are proud of their independence, and are anxious to maintain themselves. ... We are not asking for direct relief. What we are asking for is ways and means to enable our fishermen to earn a living so that they can keep themselves off the relief rolls.[28]

Robert Finn, a Liberal MP from Halifax, went further as he emphasized the privation, and distress among lobster fishermen. He acknowledged that the Minister of Fisheries had been "sympathetic," but that quick action was required to get the men ready for the spring season. Citing the federal jurisdiction over fisheries established in the British North America Act, Finn reasoned that individual fishermen were a federal responsibility and thus entitled to relief when suffering "through no fault of their own." He questioned how a government could justify prosecuting fishermen who broke the law yet refuse to support them in the face of crisis. Most pointedly, he challenged the Minister of Fisheries' remarkable claim that "fishermen are not part of the fisheries."[29] In response, the Minister insisted that despite generous federal aid in the past three years to fishermen, individual lobstermen fell under provincial responsibility and should seek help through provincial poor laws. The effort to separate fishermen from the fishery foreshadowed postwar policy trends, but it rang hollow in a context where federal fisheries' funds were simultaneously being used to build provincial roads and provide work for struggling fishers. These rhetorical contortions reflected the state's attempt to distinguish fishermen as individuals deserving provincial relief and fishermen as producers active in a national economy.

Untangling lobster fishermen from the fishery was an ongoing theme of the postwar period, as concern was less about a diminished resource and more about the fishery's associations with poverty and inefficient productivity.[30] This reorientation emerged from the work of economists who prioritized improving people's standards of living through modernization, shifting attention away from biologists who had focused on the fish. Over time, the problem of the lobster fishery was reframed as too many fishermen rather than too few lobsters.

The Economists

One of the immediate and less visible consequences of the war was the rise of economists. The presence of economists within the civil service predated the war and was connected to fundamental transformations within the discipline as it reinvented itself as a scientific rather than a descriptive tool of analysis. The economists with increased power in the postwar civil service were decidedly different from the Harvard-educated and Queen's University-based economists of a generation earlier.[31] They were all technocrats, and although they claimed political neutrality based on their expertise, they proselytized a logic based on specific values that placed the free market at the core of civil society and individuals as rational economic actors. With regard to the fisheries, historians Miriam Wright and Jennifer Hubbard have both drawn our attention to the replacement of biologists with economists in shaping policy.[32] This had tremendous consequences as economic rationale trumped social or environmental justifications.

In her splendid study of the history of the economics profession in the United States, Britain, and France, sociologist Marion Fourcade reminds us that between the 1930s and the 1960s economics itself emerged as a "technique of government (symbolized by the twin innovations of national accounting and macroeconomic modelling) and, more generally, as a tool for the exercise of public expertise." This way of seeing the world and especially abstract modelling was fundamental for new paradigms of social, economic, and ultimately environmental regulation. Although deeply political in its values of the competitive markets' efficiency and rational behaviour, economics was able to portray itself as offering unpolitical and technical expertise with "calculative rationality."[33] When Harvard University received a major donation to establish a graduate school in public administration in the mid-1930s it was not at all a foregone conclusion that it would find its home within economics; rather, this was a moment when the association between public administration and economics was in formation.[34] Historical geographer Daniel Banoub has alerted us to the fact that this mindset had already reached Newfoundland's Commission government in the 1930s, which imagined a more prosperous and efficient fishery conducted by fewer fishermen chasing a larger stock.[35] For lobster, the rise of an economic world view meant fishery economists eclipsed biologists in defining both problems and solutions. Fisheries biologist Russell Earle noted in 1947 that the Canadian federal government now "demanded" science with a mathematical focus to determine a maximum sustained yield.[36] This adoption of both economics and mathematical modelling offered the illusion of an objective reality and that policy decisions were edging away from local politics.

This change in the culture of public administration and policy coincided with a generational change in the leadership of the Department of Fisheries. W.A. Found retired as Deputy Minister in 1938 and through his former colleagues, such as R.N. Venning, could be linked back to the origins of the department. His successor British-born and Canadian-raised Donovan B. Finn was a formally trained, highly respected fish scientist who held a PhD from Cambridge. Before leaving the Department in 1946 to become the first Director for Fisheries of the United Nations Food and Agriculture Organization, Finn recruited a young academic whom he had met through the Federal Salt Fish Board, an agency established in the 1930s to regulate and subsidize fin-fish exporters.

Stewart Bates (1907–64) had no specific knowledge of fish biology or ecology but was a trained economist. Born in Scotland, he received a MA in economic science from the University of Glasgow, spent time with the Empire Marketing Board, and was an assistant lecturer at the University of Edinburgh before attending Harvard as a Commonwealth Fellow in 1934–5. There he worked with economist Joseph Schumpeter and imagined a future as an academic working in the field of public finance.[37] Instead, Bates came to Canada in 1936 originally to work as secretary to the Nova Scotia Economic Council but in 1937 was employed by the Rowell-Sirois Commission on Dominion-Provincial Relations where he wrote *The Financial History of Canadian Governments*. In 1938 he was appointed as professor of Commerce at Dalhousie University and served as a consulting economist for the Federal Salt Fish Board as an introduction to the world of fisheries.[38]

In 1942 Bates was seconded for wartime service as Special Assistant to the Deputy Minister of Fisheries. During this period he authored the influential *Report of the Canadian Atlantic Sea-Fishery* (1944) for the Nova Scotia Royal Commission on Reconstruction and Rehabilitation. Historian Miriam Wright points out that this study can be recognized as emerging from a modernization paradigm, decades before W.W. Rostow and his cohort of economists popularized this theory.[39] Bates saw that the fundamental problem with the Atlantic fishery (in contrast to the "modern" operations of the capital-intensive fishery on the Pacific) was the backward attitudes of the local population and the lack of capital and technology. The Atlantic fishery needed modern development.

Bates did not return to Dalhousie at the end of the war but was appointed by C.D. Howe, "Minister of Everything," to serve as Director General of Economic Research for the federal Department of Reconstruction and Supply before returning to Fisheries in 1947 as Deputy Minister.[40] His importance as Deputy Minister cannot be under-estimated, and historian Joseph Gough used the subtitle "Stewart Bates Stamps Departmental Mindset" to begin the postwar section in his history of the Department. Bates remained as deputy minister until December 1953 when he was appointed as president of the Central Mortgage and Housing Corporation.[41] Managing mortgages and the Canadian housing sector was as abstract as managing fish.

In his 1944 foundational study *The Report of the Canadian Atlantic Sea-Fishery*, Bates clearly understood the current importance of lobster and yet at the same time underplayed its significance for the future. Lobster did not fit neatly into the emergent modernization framework. Based on statistics from 1939, he noted that "eastern Canada as a whole was the land of the lobster, not the cod."[42] Dividing the Maritimes into eighty-nine geographic fishing districts, Bates found that only nine had no lobster fishery at all, and in another ten, lobsters generated less than 25 per cent of total fishing income. This meant that in seventy of his eighty-nine fishing districts, lobster catches provided at least 26 per cent of all fishing income. In Nova Scotia specifically, more than half the districts generated more than half their income from lobster, and in five of these, lobster represented more than 75 per cent of total fishing income.[43] Bates emphasized that lobster was central to most fishermen's income, as "the value of the commodity was so high relative to all sea fish" even though the legal open season in many districts was only around two months long. Lobster's high value and low barrier of entry made it too attractive – drawing too many participants into what was increasingly regarded as an overcrowded fishery. Bates contrasted only needing "a simple lobster pot" with the imagined future of an

Atlantic fleet of modern industrial offshore groundfish trawlers or the idealized capital-intensive Pacific salmon fishery.[44] Lobster, a fishery made only possible by the new technologies of the middle of the nineteenth century in food preservation and transportation, was now cast as the quintessential *traditional* fishery.

There is no doubt that the decentralized inshore lobster fishery was frustrating from a postwar centralized planning perspective. Although other fish species varied in price according to location, the case of lobster was extreme. In 1939 fishermen in southwest Nova Scotia received $18.75 per hundredweight, while those in Prince Edward Island earned $6.37 for the same catch.[45] The uneven pattern established in the nineteenth century and based on easy access to American consumers persisted. Moreover, as noted earlier, during the interwar period, lobster was fished more intensely. The estimated 1.5 million traps in the water in the early 1920s expanded to around two million traps by the late 1930s. As more traps were set in the water, the catch per trap declined – from twenty-three pounds to fourteen pounds – making fishing efforts and the investment in gear increasingly inefficient.[46] Efficiency was celebrated, and from his desk in Ottawa, Bates saw the lobster fishery as embedded in wasteful rural economies – communities marked by poverty and the absence of capital. Lobster's association with "casual fishing," a strategy that blended seasonal fishing with farming, forestry, and intermittent wage labour, was not understood as sustainable community-based occupational pluralism. Instead, it was cast as evidence of underdevelopment, a lack of specialization, and inadequate expertise. The fishery was part of a post-war problem that would have to be solved through "interadministrative" efforts involving "farming, forestry, fishing, public health, education, electrification, roads, and other 'amenity' capital."[47]

Bates was not alone in understanding the problem of the lobster fishery in terms of economics. One of his protégés was H. Scott Gordon (1924–2019). Born in Halifax, Gordon graduated from Dalhousie in 1944, earned an AM studying Keynesian economics from Columbia and began his doctoral work at McGill in 1946. The same year, Bates hired Gordon to work at the Fisheries Prices Support Board where he learned about fish before joining Carleton College in 1948. Gordon remained at Carleton (later University) until 1966 when he moved to the Economics Department at Indiana University, eventually holding a parallel appointment in the Department of History and Philosophy of Science until his retirement in 1988.[48]

As an economist in Ottawa with connections to Bates, it is not surprising that Gordon occasionally worked for the federal Department of

Fisheries. In 1952 the Department published a fifty-nine-page report he prepared for the Prince Edward Island Fisheries Development Committee, a group organized by the federal and local governments.[49] Like Bates, Gordon emphasized how during the 1930s there had been a dramatic increase in the number of fishermen into the lobster fishery, a fishery that had no room for expansion. In the case of Prince Edward Island, this was an increase of 50 per cent and fishermen increased their distress by incurring debts for boats and gear as their earnings fell. This resulted in a "pathetic" situation perceived of their own making that Gordon compared to a tragic fire where lives were lost in a stampede to the exit rather than by fire or smoke.[50] The economist asked, "If the size of the cake is fixed, is it not both proper and just to prevent too many from sharing it? Canada is a young and developing country. In normal times there are great opportunities for employment in new and expanding industries and while this almost certainly means a movement of population from areas like Prince Edward Island, the wealth of all will be very much greater if such movement takes place." The Prince Edward Island economy "will never be able to support a population much larger than at present without lowering the standard of living." The relocation of surplus lobster fishermen to industrial centres in Central Canada was seen as beneficial not only for reducing pressure on the fishery, but also for enriching manufacturing areas with "people who work well and efficiently but who will bring with them a knowledge of the land and the sea with their spirit and traditions which are sometimes entirely lost in industrial civilization." So, in 1952, Gordon recommended that lobster become a limited entry fishery pursued by full-time specialists.[51] At this point, except for salmon, there were no limits to access for any fishery on Canada's Atlantic Ocean.

Prince Edward Island's lobster fishery clearly influenced Gordon's seminal 1954 article "The Economic Theory of Common-Property Resource: The Fishery." This article changed the academic conversation – and eventually policy discussion – as it brought biology and economic modelling together. Gordon believed that it was possible to establish a "perfect catch," which would provide the greatest economic benefit for fishermen in terms of the investment of capital and labour. This ideal number, however, could not be achieved in an open-access commons and, therefore, individual licensing was a prerequisite. In arguing for individual transferable quotas, he concluded that the existing legal framework had emerged from concerns about overfishing but had not worked. Indeed, Gordon even questioned whether resource depletion was a genuine issue. He noted that the terminology was shifting from "fisheries conservation" to the more contemporary "fisheries management".[52] The problem, in his view, was not the fish or the environment but the way humans utilized natural resources. Biologists who had previously shaped conservation policy by focusing on fish or lobsters had treated fishermen as an "exogenous element." Gordon concluded that transforming fisheries into restricted property rights – like other forms of private property – could generate true wealth. With reference to lobsters, he noted that in the few communities where fishermen had established an informal local monopoly and kept out outsiders by customary means, these communities had both improved their incomes and reduced the amount of gear invested.[53] Echoing Keynesian economists, Gordon believed that modern states were required to manage the exploitation of nature (or resources) just as they managed other aspects of their economies.[54] A stable fishery offered economic reliability and therefore social stability. His ideas were expanded the following year by University of British Columbia economist, Anthony Scott, who insisted that resource private property rights were not only necessary but that they needed to be guaranteed to be long term, exclusive, secure, and tradable.[55] Gordon's influence reshaped both Canadian lobster policy and conservation policy more broadly as it is often associated with the "Tragedy of the Commons."[56] What has been less frequently acknowledged is that Gordon's empirical world view was formed by the Prince Edward Island lobster fishery.

Gordon's ideas were not adopted immediately, but this worldview gradually permeated the Department of Fisheries. As historian Jennifer Hubbard has shown, the general dominant explanatory frameworks for fisheries management had shifted in ways that made Gordon's approach increasingly palatable. Fisheries science in the 1930s began to adopt the concept of "maximum sustained yield" from mid-nineteenth-century German forestry. Hubbard points out that these mathematical models, less expensive than actual field investigatory research, established ideal sustainable catches for fish populations.[57] Underwater fish and lobsters, however, unlike trees, cannot be seen and counted, and population models were always abstracted. In this context, Hubbard astutely observes that "Canadian fisheries economists were not appointed to offer economic analysis; they were economic activists."[58]

Bates and Gordon pushed for a modern, industrial fishery managed by the state. It required not only the management of fish but perhaps especially in the case of lobster, the management of individual fishermen and their families.[59] The management of fishermen and their families reflected an agenda that had expanded beyond the Department of Fisheries as the inshore fishery – with lobster at its very core – was the focus of concerns about regional rural poverty. The poverty of the lobster fishermen intersected with larger federal programs connected to both the postwar welfare state and the framing of the "problem" of underdevelopment in Atlantic Canada. David Wilder, a biologist who joined the Department of Fisheries during the war and worked on lobster at the St. Andrews Biological Station, took it for granted that in working with lobster "you had to get into biology, economics and sociology."[60]

Indeed, the organization of the Department of Fisheries was changing. The 1956–7 Department Annual report noted that the Markets and Economic Service had been reorganized as the Economic Service to address "urgent economic problems that result from the application of measures of control or management, primarily in the interests of conservation."[61] Economics had severed is roots with encouraging consumer markets and was now about "control" of the fisheries. By coincidence, that same year the Department informed any report reader that the Maritime area was divided into ten different lobster fishing districts "for economic reasons," as if ecosystems, water temperatures, ice flows, and moulting patterns had had absolutely nothing to do with how the geographic boundaries of various districts had originally emerged.[62] This is not to say that the fishing districts were ever only about biology and the environment, but the physical world also mattered.

Fish economists modelling the optimum inputs of labour and capital for maximum profitability, social welfare concerns and programs, and real-world politics made for crowded bedfellows in post-war Canadian lobster fishery policy. All three streams identified the inshore fishery – with lobster as the key species – with poverty, but the possible solutions differed greatly and were often in conflict. As the welfare of fishermen was increasingly conceived as distinct and separate from the welfare of fish, income support and stabilization programs directed at fishermen were less likely to be found within the Department of Fisheries. Some residual programs continued, however, and in 1950, for example, the desperate plight of Magdalen Islands lobster fishermen and their need for emergency relief was brought again before the House of Commons.[63]

A different emergency in 1952 caused Nova Scotia to successfully lobby the federal Department of Fisheries to supply half the money for an emergency fishermen's loan program. The storm-related loss of lobster traps in the winter of 1951–2 led Nova Scotia and Ottawa to cost-share interest-free loans for up to 80 per cent of the value of traps lost to fishermen to be repaid over three years.[64] Cabinet embarked on this plan reluctantly as it expressed concerns about establishing a precedent but was assured that the criteria of an exceptional weather event would limit demand. Not surprisingly after severe storms in Spring

1953, Prince Edward Island lobster fishermen also looked to Ottawa and the Department of Fisheries for help in replacing lost gear. The timing of the Prince Edward Island storms was particularly unfortunate, as they occurred at the start of the season and meant the loss of income for the year. Since the money offered to Nova Scotia lobstermen had come in the guise of a loan, the federal government could not give money directly to fishermen in Prince Edward Island, yet duplicating the same offer was not possible. Prince Edward Island stated that it was "financially incapable" of supplying half the funds. In the end there was an 80/20 split of federal funding with the province.[65]

These two (unsatisfactory) *ad hoc* solutions resulted in the Department of Fisheries entering the lobster gear insurance business. In May 1953 the Department of Fisheries introduced the Fishermen's Indemnity Fund as a means of making available insurance for fishermen who wished to protect their vessels and traps against storm damage. Fishermen were to pay the set premium of $7.50 per hundred traps, payable in advance, and in exchange be compensated for lost traps at $1.50 per lost trap after a 25 per cent deductible. But the compensation for lost traps was set too low, as the indemnity offered was below the estimated value of $7 or $8 per trap and did not come close to covering losses. In the first year, there was virtually no uptake in the program, and by July 1953 only seven policies had been issued among the roughly 1,000 eligible lobster fishermen in the Northumberland Strait area.[66] A year later 234 fishermen and 44,927 traps were insured. The program also included boats – originally those valued between $20 and $7,500. The plan was meant to be self-financing, however, losses paid out exceeded money taken in, even with the low payout rates.[67] Although premiums were raised, the program ran at a deficit, as too few lobstermen joined and those few who did drew too many benefits. From July 1953 until the end of 1963, for instance, the fund paid out $891,176 while only collecting $231,411 in premiums.[68] There also continued to be *ad hoc* exceptional measures granted. In the aftermath of a May 1966 storm, the federal Department of Fisheries co-financed with the Government of Newfoundland compensation at $3 per trap with the condition that those who received compensation participate in the Indemnity Plan that season.[69] There were also limits to the generosity of the state. When in January 1966 a delegation of lobster fishermen from Prince Edward Island met with the federal Minister of Fisheries to request financial assistance after a failed season, it was told that this was impossible, as "the federal government had never envisaged that lobster fishermen could depend on their total annual income on a fishing season which was only two months in length."[70] The lobster fishery as a welfare program was in the midst of ending.

One path the Department of Fisheries followed to improve the lobster fishery's efficiency and productivity was the promotion of modernization through technology. This strategy was a dead end with a couple of important exceptions. In the early 1960s most lobster fishermen still used natural fibre ropes, but synthetic alternatives were adopted and seen as worth the capital investment in terms of durability and reduced maintenance. Likewise, Styrofoam and plastic buoys introduced in the 1950s and in the 1960s replaced the balsam fir and cedar models. It was not until the mid-1970s that elastic bands began to displace wooden pegs used to secure lobster claws, that fibreglass boats began to replace traditional wooden models and that some regional fishermen began to adopt plastic-coated wire traps. Fibreglass boats and manufactured wire traps were decisions that exchanged time for money. Yet in many areas with long close seasons, fishermen continued to prefer building and repairing their own wooden traps into the twenty-first century.[71]

It is hard not to think of the age of the airplane when in the late 1950s, scientists with the Department of Fisheries had some enthusiasm for aluminium lobster traps. The group of Prince Edward Island fishermen who received them on a trial basis refused to even put them in the water or discarded them after a few days, as they did not fish well and wasted the few precious weeks of the legal fishing season.[72] In the 1960s the Department of Fisheries supported research in the creation of plastic traps. The plastic traps lasted longer and appeared to use less bait, but the rationale behind the program was the hope that they would free lobster fishermen to accept "gainful employment during the off-season" and not be tied to making and repairing their own traditional traps.[73] However, their cost at $25 per unit was way beyond the capital expenditure that any lobsterman might justify. Self-made traps were primarily a matter of time. When lobster fishermen rejected plastic traps, the fisheries bureaucrats explained the lack of acceptance was the result of the "prevailing conservative attitude toward change in general" and lack of a track record. Indeed, the reaction to the plastic traps was uniformly negative, with not a single fisherman concluding that they were as efficient as traditional wooden traps.[74]

The exception to the adoption of a modern, industrialized lobster fishery was the development of an offshore lobster fishery in the waters between the 50 and 200 nautical mile boundaries. In 1965 the Department of Fisheries permitted an experimental offshore lobster fishery off the Atlantic coast. A 1970 attempt to formally open this fishery met with serious opposition from inshore fishermen, who feared this might threaten breeding stocks.[75] Nevertheless the offshore fishery was institutionalized in 1971 as Lobster Fishing Area 41 and, it was justified as a form of compensation for longline swordfish fishermen who had lost their livelihoods when that fishery was closed due to high mercury levels in swordfish. Unlike the adjacent inshore lobster fishery, which is characterized by independent commodity producers, LFA 41 has been fished corporately and characterized by a geographic monopoly, a set quota, and gigantic trawl lines holding one hundred traps. Had Stewart Bates lived to see this, he likely would have been pleased.[76]

On the other hand, Bates and the Department of Fisheries did not support Fishermen's Benefits within Unemployment Insurance, the most important program to raise inshore lobster fishermen out of poverty. While income supplements improved the standard of living, unemployment insurance was held responsible for encouraging overcapacity and obstructing efforts for an efficient and productive fishery. Although Unemployment Insurance was introduced to most wage workers in the 1940s and was regarded as a cornerstone of the welfare state, self-employed fishermen as "co-adventurers" and not wage workers were originally exempted from the program. Coverage was eventually extended to other excluded seasonal workers, such as in forestry, and after 1950 some regional politicians, led by Newfoundlanders, actively lobbied for the extension of benefits to include self-employed fishermen and succeeded with a special amendment to the 1956 Unemployment Insurance Act. The new program began in April 1957, with the first payments made in January 1958. Fishermen's benefits were based on the net value of a week's catch, which was generally defined as the gross value less 25 per cent operating costs with a bottom minimum threshold to be considered an "insurable week." The 1985–6 Forget Commission on Unemployment Insurance critically described this amendment as "essentially a political decision, motivated by social rather than economic considerations."[77]

In June 1952 the Department of Fisheries had approached economist Scott Gordon to make recommendations that might be alternatives to Unemployment Insurance.[78] Nothing came of this overture, but the Department views were reflected in the 1953 Walsh Report that opposed the extension of Unemployment Insurance to self-employed fishermen. This report focused exclusively on Newfoundland, and the vision of a technologically advanced fishery with a smaller workforce prevailed.[79] The introduction of Unemployment Insurance provided an income supplement to poor fishermen and inadvertently encouraged precarious fishermen to remain in the very fishery that economists in the Department of Fisheries were trying to reduce in size. Like the actuarial numbers in the trap indemnity programs, the numbers did not work out to be self-supporting, and it was estimated that fishermen's benefits cost the Unemployment Insurance fund in 1960–1 alone $10.4 million. Unemployment Insurance raised the income of inshore fishermen through regional income transfers – not from any improvement in productivity.[80] In 1962 the Gill Committee's report into the Unemployment Insurance Act recommended that the "economic and social problems" associated with fishermen and the fishing industry were not suited for the program and that the Department of Fisheries administer a separate plan. It should not be surprising that this was politically unpopular and not taken up by the Department.[81] Although Unemployment Insurance was not distributed through the Department of Fisheries, it altered the lobster fishery as much as any conservation policy directed towards fish.

The Department of Fisheries' fear that income support through unemployment insurance would keep more people in the fishery was correct. Post-war social goals were to maximize male employment, and the lobster fishery inadvertently became part of the welfare support system.[82] Rural outmigration of young people from high birthrate areas in the Atlantic region characterized the 1950s and 1960s, but the number of people engaged in the fishery actually grew.[83] As lobster fishing was often an occupation of last resort, unemployment benefits may have made it more attractive for those who qualified. A Department of Fisheries study from 1961 revealed that about half, or 52 per cent, of all lobster fishermen received Unemployment Insurance payouts that year.[84]

That approximately half of all lobster fishermen did not qualify reflects how complex the rules were. As payments were made on the value of the catch rather than hours worked, benefits were unequally distributed. Lobster was a high-value fishery, so this meant higher payments to Nova Scotia lobster fishermen than Newfoundlanders pursuing groundfish. There were also issues about accessibility, as in certain districts the legal season could be as short as ten weeks and this intense period of fishing might be interrupted by bad weather.[85] The lengthy close periods meant that lobster fishing could never be anything more than seasonal work, and in the extension of Unemployment Insurance eligibility to lobster fishermen, the federal government played a direct role in modernizing occupational pluralism – an issue at the same time being cast as a traditional "problem" identified with inefficiency and poverty. Sociologist Derek Johnson, in his study of an Acadian fishing village on the Shippegan Peninsula, argued that Unemployment Insurance benefits paid to lobster fishermen offered

continuity with the traditional survival strategies that had been provided earlier by merchants in carrying fishing families financially during the off season.[86]

But this was exactly why there had been opposition from the economists within fisheries. Fishing for unemployment "stamps" rather than
for lobster gave new power to the legal buyers (who essentially controlled the credentials for qualification for benefits) and transformed
individual fishermen's strategy around how much should be caught.
During the first years of the Unemployment Insurance program, the
lobster buyer had to insure both the captain and a boat's hired men.
In 1958 if a fisherman earned $50, he contributed 16¢, while the lob-
ster buyer had to pay 32¢ in premiums. Recipients with no dependents
received $23 a week and were eligible for one week of benefits for every
two weeks worked after the first of April.[87] Rural consumers or tourists looking to buy lobsters might not be successful, as in a 1962 case
when fishermen with lobster in their possession refused to sell to a
"wayfarer," as he "could not offer them any unemployment stamps."[88]
The recognition that social, cultural, community, and political decisions
rather than purely market-driven or ecological factors shaped policy
would eventually be encapsulated in the concept of a "social fishery."
This approach prioritized maximizing employment and sustaining
coastal communities over efficiency or ecological considerations. Decisions about regulation and government policy were taken with the fisherman rather than the lobster in mind.

Certainly, it was with the fishermen in mind that the director of the
St. Francis Xavier University Extension Department lobbied in 1956 to
close the residual canning sector along the Guysborough Shore. The
proposal was to increase the size limit, thereby directing lobsters into
the higher-value live lobster market stream.[89] The size limit here had
nothing to do with the conservation of the lobster stock and everything to do with raising the standard of living for inshore fishermen.
Ten years later, the same extension department explained that the
"problem" of the inshore fishery on the Guysborough Shore was not
the collapse of the lobster stock, although catches were falling, but the
poverty of fishermen. Electricity, washing machines, indoor plumbing,
and central heat were counted alongside lobster landings.[90] Echoing the
consensus emerging in the Department of Fisheries, a report from the
Extension Department of St. Francis Xavier University recommended
a lobster fishery with trap limits restricted to those over twenty years
of age who had permanent residency in the community. The age limit
had social goals of keeping boys in high school. Although 70 per cent of
the local fishing income came from lobster, the report recommended a

year-long open season to encourage specialization in either the lobster or the groundfish sector.[91]

A proposal to use lobster regulations to reduce high school dropout rates and raise education levels rather than conserve a natural resource stood in stark contrast to the concurrent push to create a modern economically productive and efficient fishery. It is important to realize that both policy avenues were being pursued simultaneously within the federal government; the economic fishery, based in the Department of Fisheries, focused on the capital-intensive offshore, while the inshore social fishery found support in other departments and from regional politicians. Despite their contradictions, these strategies could overlap, as seen in attempts to modernize the lobster fishery. Initiatives to increase fishermen's income and improve the efficiency of their efforts aligned with both agendas. Historian Jennifer Hubbard has alerted us to the presence of underdevelopment rhetoric within the Department of Fisheries, and this language of community development resonated with the social fishery.[92]

In 1961 the Assistant Director of the Economic Service of the Department of Fisheries economist J.B. Rutherford led a study with department economist H.C. Frick and biologist David Wilder published six years later as *An Economic Appraisal of the Canadian Lobster Fishery*.[93] Rutherford was the first MA graduate from the Saskatchewan College of Agriculture in 1931 and went on to become the Chief of the Agricultural Division, Dominion Bureau of Statistics before joining the Department of Fisheries.[94] He was heavily influenced by Scott Gordon's conclusion that an open-access fishery was at odds with maximum efficiency. Rutherford's study set the future direction of lobster regulation as it outlined an extreme argument for deregulation, and its delayed publication can only be understood in a greater context of fisheries politics and the challenges of political acceptance this report posed.

In *An Economic Appraisal of the Canadian Lobster Fishery*, lobster fishermen were framed as micro enterprises hampered by inefficient and inconsistent regulations around open seasons, licensing, gear, and size but who were also subsidized with indirect assistance through "money losing programs" such as the Fishermen's Indemnity Plan. Anyone could get a licence to fish lobster in a specific district for 25¢ ($2 by 1968), with Quebec fishermen charged through their provincial authority the fee of 25¢ plus 1¢ per trap fished with a limit of 300 lobster pots. Open seasons ranged in extremes from 222 days in southern New Brunswick to a mere sixty-one days on the eastern Atlantic coast of Cape Breton and the Magdalen Islands. The gear was itself frozen in nineteenth-century technology while Americans experimented with offshore draggers and diving. Too much time and capital investment were being put into inefficient and immobile gear, while the short open seasons restricted lobstering to a "small-scale, part-time fishery."[95] Throughout this important report, the reader gets hints that the lobster fishery was the most important inshore fishery keeping communities alive, but its foundational role was not given the same importance as making the fishery more efficient.

Social factors were also dismissed in Rutherford's report. Department of Fisheries historian Joseph Gough recounted the frustration economists had with the lucrative lobster fishery in southwest Nova Scotia. Here the long winter fishery opened around the first of December and ended at the end of May. Weather and market conditions meant that there was intense activity in December, but fishing did not really resume until March or April through May. The December fishery was deemed inefficient, but fishermen opposed changes as they wanted money for Christmas.[96] Not only was the legitimacy of this objective dismissed but so were the good market conditions connected to higher prices and seasonal consumer spending, generally decent weather for fishing, and the superior condition of the lobsters for shipping in December with their full hard winter shells.

In outlining an argument for deregulation and noting the cost of enforcement and the efficiency of the free market, the study concluded that solving the lobster question was a "political" problem. To make the lobster fishery more productive and to increase the income of those fishing, there needed to be a drastic reduction in the number of fishermen. The predicament was cast again as not too few lobsters but too many fishermen. Neither lobsters nor fishermen needed to be conserved, but both should be managed. *An Economic Appraisal of the Lobster Industry* attempted to build common ground with fishermen in highlighting their "resentment" towards "moonlighters," those who had other occupations but took time off or used their holidays or spare time to fish during the short legal season. Rather than highlighting the lobster fishery's accessibility – its relatively low capital requirements and its integration with long-standing household strategies of occupational pluralism – the report redirected attention to broader regional social and economic challenges. Chief among these were the limited alterative occupational opportunities and the perceived lack of labour mobility in Atlantic Canada. The lobster fishery was thus positioned not as a viable local livelihood, but as symptomatic of deeper structural issues.[97] Models of progress required that both capital and labour be mobile and move to where they were most productive and efficient, a strategy incompatible with the dispersed, seasonal, and community-rooted nature of the inshore lobster fishery.

While the report was sitting unpublished, its assumptions were dispersed in other department venues. The report's recommendations were foreshadowed by the department's F.E. Popper, who in 1956 worked for the Prince Edward Island Fisheries Development Committee and recommended the allocation of property rights in the lobster fishery to community-owned companies. Any access disputes would be addressed through private civil action. W.C. Mackenzie, the Department's director of economics in 1965 referred to "the development of lobster ranching" as if lobsters were privately own chattel like cows.[98] He had made this comparison explicit in a 1960 editorial that noted that there was "no need for the Canadian state to regulate the killing of cattle ... Farmers have control over their cattle, through private ownership, while fishermen have not control of the lobster on which they depend."[99]

By the time the Rutherford, Frick, and Wilder report was finally published, its dramatic recommendation to bring lobsters under "single management as a private enterprise or as a state monopoly" had become a familiar argument. What sociologist Patricia Marchak later termed as a "state property regime" would restrict the entry of both labour and capital.[100] The 1967 report posed the option that resource management could be offered to community companies or cooperatives but endorsed the idea of a free market where, after a five-year transition period when no new licences were granted, thereafter, licences were only to be issued by the state after "competitive bidding."[101] If, as sociologist John Flint argues, the postwar period had been marked by an effort to manage fisheries for community and regional development, we see here also the use of market mechanisms.[102]

Although the report did not set a precise regulatory framework, the speed of change capitalized on a moment of crisis, as lobster landings crashed. The conservation success of existing regulations could be dismissed as catches were in freefall. In 1960 the Maritimes landed 51.5 million pounds of lobster, but this number declined, reaching a low of 31.3 million pounds in 1974, among the lowest landings in nearly four decades.[103] This stock crisis created the conditions under which drastically different regulatory approaches could be entertained.

In Quebec, where the lobster fishery was managed by the province, limits on the number of traps a lobster fisherman set were in place by the 1960s, but these limits did not appear to effect "exploitation rates." The issue of trap limits was taken up by the Economics Branch of the federal Department of Fisheries in the 1965 season when information was collected from lobster fishermen in Lismore, Little Harbour, and Big Island in eastern Pictou County. Since 1941 lobster fishermen in Lismore had set an informal trap limit and restricted entry so that the number of traps used by each boat fell from 500 to 250. They protected "their" area by what fisheries bureaucrats obliquely referred to as "various methods of persuasion," only welcoming local fishermen's sons. With less efforts and gear, their catch did not change. In the late 1950s fishermen in nearby Little Harbour began to reduce the number of traps they also fished, catching the same number of lobsters with fewer traps. Lobster fishermen on Big Island made no change in their behaviour. In the 1965 season lobster fishermen in Lismore and Little Harbour earned about $3,300 and those at Big Island $2,100 for two months of fishing and between three and five weeks of trap repair. Most of the fishermen engaged in "non-lobstering" activities for the rest of the year that only accounted for less than half of their annual incomes.[104] The findings of this largely observational study justified Fisheries Minister Hédard-J. Robichaud's March 1966 announcement to introduce an experimental trap limit in Lobster Fishing District 8 (the Northumberland Strait).[105] The following year it was extended to the rest of the southern Gulf of St. Lawrence fishery. The direction mirrored the Rutherford et al. study but was tempered by politics, with the result of what historian Joseph Gough described as a "milder approach."[106] Just as earlier conservation and enforcement were limited by community-level politics, efforts to move to a "state property regime" were constrained by electoral politics.

In June 1967 District 8 again became the flagship for new policy when licence restrictions were introduced – no fisherman in the Northumberland Strait could get a licence to legally fish for lobsters if he had not held one for the 1966 season. This policy was extended in February 1968 to all lobster fishing boats in the Maritimes. Newfoundland came under this regime in the mid-1970s.[107] The legal right to fish lobster was frozen, and with no advanced warning fishermen who did not hold a licence in 1967 were permanently shut out. Trap limits were also introduced, with different maximums set for different districts. In the 1968 season, serially numbered metal tags were provided by the Department of Fisheries and were to be affixed to each lobster trap. Untagged traps were deemed illegal and could be confiscated or cut from their ropes.

Licence and trap limitations were only the first step. In January 1969 an individual's access to a lobster licence, which had cost $2, was transferred from a personal right to the privilege of a specific vessel to engage in the fishery. Lobster licences stayed with the boat, and a department statement described the boat as "the passport to the lobster fishery." A month later, additional regulations were released. In the first place, a provision that had permitted two licensed fishermen who jointly owned and operated a single boat to fish additional traps was removed. The boat was all that mattered. But more importantly – the lobster fishery was divided up into different classes of fishermen with the stated objective of distinguishing "bona fide fishermen" and others who caught lobsters on a casual, weekend, or part-time basis, as one of the many components in putting together a livelihood. Class A licences were issued to those considered "real" fishermen and who were not employed elsewhere. Their boats were issued licences that could be transferred to the new owners if a boat was sold. Fishermen who fished, depending on the area, fewer than one hundred, seventy-five, or fifty traps in 1968 were assumed to be part-timers with employment elsewhere. They were restricted to a limited number of traps going forward and were assigned Class B boat licences. As part-timers their vessel licences would be extinguished when they retired.[108]

But even this new regulatory direction, reflecting economic goals was not without social policy. Fishermen in the Gulf, who had only relatively recently freed themselves from the control of buyers and processing companies, feared that with licence limits, corporate interests would soon have control of the boats again. In April 1969, restrictions were placed on multiple licensing, no person or company could license more fishing vessels than they had registered in 1968. In addition, by custom, lobster fishermen did not fish Sundays, but Sunday fishing only became illegal in 1968 as customary practice was reified in law to curtail the activities of "moonlighters" who may have been employed elsewhere and only fished on Sunday.[109]

Apart from limitations on drift net Atlantic salmon operations introduced in the 1930s, lobster became the first limited entry fishery in Canada. This precedent spread with remarkable speed to other species of fish. The Atlantic lobster plan was mirrored on the West Coast by the September 1968 "Davis Plan" for the "fleet rationalization" of the salmon fishery, which involved limiting the number of fishers, creating different kinds of licences, and unlike the lobster fishery in 1968, the state immediately began buying back vessels and retiring other licences.[110] It was widely understood that the Department of Fisheries' thinking around limited entry lobster fishery emerged from the campaign of Pacific salmon fishers through British Columbia's United Fishermen and Allied Workers' Union, which had been pushing for licence limitations since 1948 in the hope of increasing incomes.[111] On the Atlantic coast, in 1971 limited-entry licensing began for the salmon fishery and herring-seine fishing. In 1972 limited-entry licensing expanded to Bay of Fundy scallops, offshore scallops in 1973, and groundfish beginning in 1974. By 1976 all commercial fisheries on the Atlantic were restricted to licensed vessels. The ocean commons was closed.

It is impossible not to make a connection between Canada's closing of the lobster commons in 1968 and Garrett Hardin's influential publication of the "Tragedy of the Commons" the very same year. A PhD in microbiology, Hardin's work focused on human overpopulation, and unlike the earlier economists who shaped Canadian lobster policy, Hardin was a critic of the welfare state. Many of his subsequent critics have pointed out that he did not differentiate between common property with mixed informal and formal practices of community regulation like the lobster fishery, and a completely open-access resource.[112]

The timing of the closing of the lobster fishery reflected a moment with falling catches, rising prices, the election of a majority government in Ottawa, and the victory of neoliberal economists in the Department of Fisheries. The centralization and uniformity promoted by Keynesians that emphasized predictability and abstraction evolved into neoliberalism and the belief that allocating individual property rights was the only viable strategy to prevent overfishing and through efficiency raise the standard of living for those involved.[113] This fundamentally transformed the lobster fishery – ending the resource commons and privatizing access as boat licences soon took on inconceivable monetary value. The privatization of the right to fish lobsters reflected a shift in the world view that neoliberalism brought.

Conclusion

Fishermen, their families, and their communities were as integral to the fisheries as fish themselves. While federal concern for lobster stock preservation dated back to the 1870s, it was consistently moderated by political pressures to sustain livelihoods and reward partisan allies. Conservation and sustainability were understood as encompassing an ecosystem that included both people and fish. Legislative, judicial, and political responses to crises in the fishery often reflected a range of sympathetic state responses. When districts were defined to complement other local economic opportunities, when seasons were extended as a form of community relief, when there was leniency in the enforcement of the law, when public works were constructed out of Fisheries budgets, or when appropriations were granted to assist with storm-damaged goods, there was a justification similar to that expressed in the rhetoric around the deserving poor. Lobster fishermen were portrayed as hard-working men trying to support their families but who had encountered hard times "through no fault of their own." The

Department of Fisheries' lobster conservation policy was about providing support and assistance to fishermen and their communities.

In the postwar period, support for lobster fishermen increasingly came from federal departments beyond the Department of Fisheries. Income stability for lobster fishermen and the influence of economists within the Department of Fisheries meant that various factions within the federal state were pulling policy in competing and contradictory directions. The economic fishery and the social fishery coexisted, but not without tension as the imperatives of efficiency and rationalization clashed with political realities. Income support programs for fishing families and communities such as Unemployment Insurance and Old Age Security meant that government appropriations were no longer coming from a single departmental budget. Social policy reshaped the lobster question again in the postwar period with the clear emphasis on fishers and their communities, albeit communities that would be smaller in size and with higher standards of living.

Conclusions and Epilogue: Contested Catch

The title *Contested Catch* was deliberately chosen to evoke contemporary struggles within the commercial lobster fishery and to encourage readers to see the direct connections between the past and the present. Today, as a century ago, the lobster fishery remains deeply political. Yet contemporary debates and policy responses often overlook the fishery's complex history. Without an understanding of the first hundred years of the commercial fishery, much of the post-1970 period of limited entry licencing is difficult to grasp. The longstanding tension between the legitimacy of the distant federal state and the intensely local nature of the fishery persists, even as lobster has become a multibillion-dollar industry. Local understandings of rights – both formal and informal – continue to shape access and participation. While licences after the 1970s were a "passport" to the fishery, this right was layered upon other, older notions of placed-based access claims.

Although lobster had been used by Indigenous peoples and early settlers, it did not become a valued commodity and an important commercial fishery nor fished intensively until technology and transportation infrastructure solved the obstacle of getting a safe product to its market. After exploiting the abundant stock in the northeastern United States, in the 1870s American-led lobster canning expanded in a northeasterly direction. By the end of the nineteenth century, lobster rivaled groundfish as the most valuable fishery on Canada's Atlantic coast. It surpassed fin-fish in the number of people employed at least part of the year and had an enduring and widespread importance in most coastal communities. Lobster was and remains the backbone of the inshore fishery and as stated before, eastern Canada was "the land of lobster, not cod."[1] Both the packing sector, which remained dominant until just before the Second World War, and the more lucrative live market trade were intimately tied to industrial capitalism but only impacted by industrialization in very limited ways. The methods used to catch, process, and ship lobsters did not change greatly until after the 1960s.

The archival research that informs this book ends with the closure of the ocean commons for lobster in the late 1960s in the Maritimes and mid-1970s for Newfoundland. In thinking about the conclusion, it became obvious that some preliminary account of the past fifty years was necessary for the reader to understand why this pre-1970 period continues to matter and shape present-day politics. What follows is a short narrative account that acts as a bridge to the present.

Between 1873 and 1970, the meaning of "the lobster question," how it was defined, and how it was to be solved changed many times. A constant, however, was the maintenance of a delicate balance between the conflicting political, social, and economic interests of local fishermen, lobsters, and the commercial fishery. It should hardly be surprising, therefore, that this dynamic continues to the present. What was unexpected for everyone; the fishermen, the processors, and what became the Department of Fisheries and Oceans in 1979, was the health of the sector, and after 2000, the upsurge of lobster's biomass, with the simultaneous collapse of the rest of the inshore and offshore fishery. In the mid-1980s, as lobster catches climbed to levels unknown in a century, fishery scientists had anticipated a collapse. At the same time there was misplaced confidence that groundfish stock would continue to grow. Of course, in the late 1980s groundfish populations could no longer be sustained, leading to drastic quota cuts, plant closures, corporate bankruptcies, and ultimately the 1992 northern cod moratorium. With 40,000 people put out of work, it was the largest single lay-off in Canadian history.[2] The modern offshore industrial fin-fish fishery came to a screeching stop.

By 2017 lobster accounted for 77 per cent of the landed value of all fisheries combined for Prince Edward Island, 53 per cent for New Brunswick, and 57 per cent in Nova Scotia.[3] It was by far Canada's most valuable species, accounting for over 38 per cent of the value of all commercial landings. Moreover, in the Maritimes alone, it was fished out of some 300 communities. The surviving Atlantic fishery for many places was likely to be a lobster fishery. In Lobster Fishing District 34 (roughly Barrington, Nova Scotia, to Digby Neck) the landings from 1980 to 2016 increased by 600 per cent. Lobster was the success story, and the growth of lobster stocks benefited at least in part from the loss of key predators such as cod, haddock, halibut, and monkfish.[4] Today Canada lands roughly three times the volume of the US catch. What was once known as a "poor people's fishery" – a livelihood that had generated great wealth for a few packers and held political weight due to the large number of voters involved – has been transformed into a highly capitalized, potentially lucrative enterprise, restricted to a select group of individuals.

The collapse of the offshore corporate fin-fish fishery meant that lobster was more important than ever for the survival of coastal communities. A 2012 report prepared for the three Maritime provincial governments described lobster as "lifeblood."[5] Between 1972 and 1999, the volume of catches tripled, while the value of lobster caught increased fifteen times over, from $37 million in 1972 to $537 million in 1999. If you adjust this number to 1999 dollars, we still get a nearly 350 per cent increase. By 2023 Canadian lobster exports were valued at $3.5 billion.[6] Lobster was sold as both live and processed, although this now meant frozen in cans, whole in brine, and, starting in the 1980s, thanks to cryogenic technology, the development of a new product in freezing raw tails. It remained labour intensive, and the seasonal employment factories once provided to local women and children were now more likely to be staffed by temporary foreign workers.[7] Canadian lobster remained an export business with the United States, Europe, and more recently China serving as the most important markets.

In the 1970s, however, there were few reasons to be optimistic about the future of the lobster fishery. Despite the introduction of the limited entry in the late 1960s, catches continued to decline until 1975. While the number of licensed vessels fell between 1968 and 1972, the number of traps in the water increased significantly.[8] The crisis in the lobster fishery persisted, even as fisheries in general dropped from the federal government's agenda.

In 1969 fisheries no longer merited its own department and merged briefly into the Department of Fisheries and Forestry (1969–71). After 1971 fisheries received a specific Minister of State in the new federal Department of the Environment (1971–9). Environmental stewardship was an important political issue, but efforts to balance the protection of the environment with the competing demands of economic growth in a national economy, or at least the local employment dependent on natural resources, remained an irreconcilable challenge. In March 1979, just months before the federal Liberal government lost power, the Department of Fisheries and Oceans was established to capitalize on Minister Roméo LeBlanc's popularity among the Atlantic inshore fishermen.[9]

As Minister responsible for the fisheries, LeBlanc established the Lobster Fishery Task Force in 1974. Composed entirely of government employees it was chaired by economist Gordon DeWolf, then working for the Department of Regional Economic Expansion. DeWolf's ideas were already familiar, as he had recently published *The Lobster Fishery of the Maritime Provinces: Economic Effects of Regulation* while with the Department of Fisheries. In it, he expressed scepticism that any government regulations had helped to conserve the lobster stock and argued that "Conservation, as an end in itself, is not a valid objective for fishery management. Management goals must be related to the users of the resource, whether they be producers or consumers."[10] His Task Force report, tabled in March 1975, echoed the Rutherford report of almost ten years earlier. There were still too many fishermen, and this kept incomes low. The report called for a more professionalized fishery, recommending further trap limit reductions, the extension of trap limits to Newfoundland, and the exclusion of those not deemed to be "legitimate" or full-time fisherman. It also suggested transferring the licence from the vessel to the fisherman.[11]

On New Year's Eve 1975 Minister LeBlanc announced that to eliminate "moonlighters" licences would be frozen immediately for both fishermen and vessels. Within three weeks individuals with full-time year-round employment would be notified and lose their "privilege to fish for lobster," subject to appeal.[12] There was a fuller announcement the following September with three kinds of licences to be recognized, the first two types remaining largely unchanged. Class A licences were restricted to those who were full-time fishermen dependent on the lobster fishery. Class B licences were assigned to those who had regular employment elsewhere but had fished lobster consistently. These fishers would only be permitted to fish 30 per cent of the district trap limit, and their licences were nontransferable and normally expired with retirement. Class C licences were given to those who had full-time, stable employment elsewhere and they were only temporary licences, giving their holders a two-year window to sell their boat and gear. District advisory committees would assist the department in determining where individual licence holders would fall. These categories received general support among full-time fishermen. Among the groups being squeezed out of the lobster fishery in the effort to reduce the number of licences were Indigenous fishermen who had been lobstering on a small commercial scale. They did not fit easily into the new model.[13]

To reduce the number of lobster fishers further, in 1977 the federal government began a pilot project to buy back licences in Prince Edward Island, and in 1978 this was extended to the entire Maritime region. This program eliminated 22 per cent of all lobster licences in the Maritimes and reflected a broader trend between 1964 and 1990, during which the number of licences in New Brunswick, Prince Edward Island, and Nova Scotia fell by close to 70 per cent.[14] Lobster licences, now a restricted commodity, began to acquire monetary value, though in the late 1970s it was still modest. In New Brunswick and Nova Scotia, around 1,600 lobster licences were retired at a total cost of $5 million, averaging of $3,125 per licence. Compensation was set at the average value of individual landings in the previous three years, with a cap of $6,000. For those who had paid just $2 for an annual licence a decade earlier, this could appear to be a windfall.[15] By 1983 nearly 3,000 fewer licences existed than there had been when the limited-entry fishery was introduced. One of the unintended consequences of the buyback program was the growing and now entrenchment of the common understanding that licences were a form of property. Technically lobster licences were privileges and not property, a position upheld by the courts until 2008.[16]

With the objective of reducing the number of fishermen fishing lobster, in May 1976 Minister LeBlanc released his *Policy for Canada's Commercial Fisheries*. This framework institutionalized the idea of the social fishery and asserted that "net social benefits" should be the guide in establishing fishery management policy. The policy claimed that "a fundamental restructuring of the fishing industry is inevitable. It will come about either in an orderly fashion under government auspices or through the operation of inexorable economic and social forces."[17] Two themes in the policy are worth note for the lobster fishery. First is the almost complete focus on the groundfish fishery, which was facing a crisis more serious than what was happening with lobster. Overall, for the Atlantic Provinces and Quebec, in 1973 there were slightly more fishermen pursuing groundfish than lobsters (21,600 in the groundfish fishery compared to 20,400 fishing for lobsters). However, in the Maritime Provinces alone, there were 13,500 lobster fishermen compared to only 7,400 fishers pursuing groundfish.[18] As the fisheries people in Ottawa had their eye to the modern industrial offshore fleet with its onshore processing factories, they continued to underplay the reality that lobster was the fishery that both employed the most people and generated the most income, at least in the Maritimes. Without being explicit, the policy imagined a future when lobster was less important. Secondly, in explaining the "causes of distress" in the groundfish fishery, the first reason was titled the "Tragedy of the Commons" where "individuals must be checked" because of the "conflict between individual interests and the collective interests."[19] The "Tragedy of the Commons" model was now enshrined as government common sense.

The timing of the effort to reduce lobster licences and institute a social fishery was significant. The idea of the professionalization of inshore fishermen coincided with the Canadian government's 1977 claim of the 200-mile offshore exclusive economic zone and borrowing programs to encourage related technological investments. Sociologists John

L. McMullan and David C. Perrier note the irony that a policy direction aimed at rationalizing capacity and increasing fishermen's incomes "ultimately created a severe debt-dependent situation for fishers."[20]

Timing also was crucial for the second major social fishery policy initiative issued just before the 1979 federal election in what was regarded at the time as a blatant (if unsuccessful) effort to attract inshore fisheries households to the Liberal Party.[21] Fisheries Minister Roméo LeBlanc expressed his open support for inshore fishermen with the Atlantic fleet separation policy, which restricted the inshore lobster fishery to owner-operator boats.[22] The Atlantic fleet separation policy excluded corporations from holding new fishing licences on vessels less than sixty-five feet and was intended to keep the most important inshore fishery – lobster – out of corporate hands and preserve rural communities, as other marine resource opportunities collapsed. Fishers and processers were to be kept distinct. This was followed in 1989 by the requirement that the Atlantic inshore fishery be "owner-operated," meaning the fishermen who held the licence or quota – with very few exceptions – must do the actual fishing themselves. It is striking that these interventions into the lobster fishery took place at a time when the dominant ideological framework of much of the Western world pushed towards deregulation and the confidence in global free-market capitalism.[23] The social fishery connected to lobster was going in the opposite direction, suggesting that local politics and context still mattered. The decision to keep corporations out of the lobster fishery reminds us that social objectives did not disappear and politics, state sympathy, and even the traditional moral economy continued to influence fishing policy even as neoliberalism and market forces that emphasized privatization took on more influence.[24]

This assumed a change in mindset that did not completely happen. Lobstermen did not completely accept that the right to catch lobster was a state-awarded privilege but mapped new regulations onto local informal practices. As such the lobster fishery continued to be a "law and order problem," and issues around trap limits and compulsory tagged traps were especially contentious as they went against the traditional idea of the ocean being some kind of a commons.[25] Residual notions of older rights continue to this day, and many people believe it is only reasonable that anyone connected to a place should be able to set a trap to catch lobsters for their own personal consumption.

Opposition to lobster regulations took both formal political routes and, at times, direct action. In December 1976, 250 lobstermen and their supporters demonstrated in Shelburne, Nova Scotia, against the issuing of offshore lobster licences. During the protest, the local buyer

Continental Seafoods Ltd., who was purchasing offshore lobsters, had windows and gear smashed, property destroyed, and there were threats to burn and sink boats.[26] Volatility resurfaced in Spring 1982 when the Department of Fisheries and Oceans began enforcing a 375 trap limit for District 4, (soon to be renamed Lobster Fishing Area 34), a restriction initially introduced ten years earlier for Southwest Nova Scotia. As it seized traps or cut ropes for pots it deemed to be illegal, tempers flared. In early May about a hundred fishermen marched on the regional Department office in Yarmouth where they were met with RCMP officers in riot gear. While two additional demonstrations were peaceful, tensions escalated dramatically on 11 May 1982, when thirty fishing boats from Woods Harbour chased two DFO leased patrol vessels into West Pubnico Harbour. Concerned for safety, the RCMP ordered about a dozen fishery officers off their boats. One vessel was burned at the wharf and the other towed out, repeatedly rammed, and sunk. The result was fifty-two charges against thirteen fishermen, including nine counts of piracy.[27] In the aftermath of the Pubnico riot, enforcement efforts appeared to ease, likely due to fears of a political backlash. By 1986, it was estimated that about 20 per cent of fishermen in District 4A were setting as many as 700 traps, almost twice the legal limit.[28] As ever, enforcement of the regulations had a social context.

Settler-Indigenous Conflict

After 1990, the most significant social and political framework shaping the lobster fishery was Indigenous sovereignty and related judicial decisions. The landmark 1990 *Sparrow* decision of the Supreme Court of Canada recognized Indigenous access to a communal food, social, and ceremonial (FSC) fishery. It also emphasized the federal government's obligation to consult with Indigenous groups if these fishing rights were to be restricted. Under FSC provisions, lobsters could not be legally bought, sold, traded, or bartered.[29] The commercial fishery that developed in the second half of the nineteenth century had employed Indigenous peoples as factory workers but, as noted earlier, had generally excluded them from the fishery. L'nui Mnikuk (Lennox Island), Esgenoôpetitj (Burnt Church), and Puksaqte'kne'katik (Pictou Landing) had at least some commercial fishermen in the community, although fishers from Esgenoôpetitj were among those who sold their licences in the late 1970s in the buy-back programs.[30]

Following *Sparrow*, the federal government attempted to mould the Indigenous fishery into a form that would be compatible with the property-based restricted access of the commercial fishery. In 1992 the federal government introduced its first Aboriginal Fisheries Strategy and began the Allocation Transfer Program (1994) whereby retired commercial lobster licences were reallocated to First Nations as communal licences.[31]

The situation grew significantly more complex and unsettled after the Supreme Court of Canada's 1999 *Marshall* decision. By that time, Mi'kmaw Donald Marshall Jr. of Membertou was already a well-known figure across Canada as one of the most prominent victims of wrongful conviction – a case in which systemic racism played a central part. In the early 1990s, Marshall once again faced the courts, this time charged with fishing violations. He contested the charges under the Treaty of 1752 but over the six years the case took to come before the Supreme Court, the legal focus shifted to the Peace and Friendship Treaties of 1760 and 1761. These treaties, unlike later land-cession treaties, were interpreted by the Court as affirming not only the rights of the Mi'kmaq and Wolastoqiyik (Maliseet) to hunt, fish, and gather but also to earn a "moderate livelihood" by trading resources. The Court's ruling *R v. Marshall* (1999) confirmed that these First Nations in Nova Scotia, New Brunswick, Prince Edward Island, and the Gaspé Peninsula had the treaty right to catch and sell fish and other natural resources. A short eight weeks after the initial decision, the court added the highly unusual amendment, known as *Marshall II*, to define "gathering" and clarified that treaty rights were not without restrictions and the federal government had the right to impose limitations for purposes such as conservation, always with the obligation to consult on treaty-related resource rights. In the twenty-five years that followed, the vitally important moderate livelihood part of the ruling remains undefined.[32]

Indigenous lobster fishing began immediately, as did the often-violent confrontations between Mi'kmaq and commercial fishermen.[33] Dangerous encounters, such as off Esgenoôpetitj/Burnt Church, were place-specific and followed a pattern of earlier violence occurring most regularly in bounded waters such as Miramichi Bay. A longstanding tradition of local violence and "direct action" now combined with the colonial reality of structural and interpersonal racism directed at Indigenous people. In this volatile aftermath, the DFO was clearly unprepared and responded by bringing the full coercive power of the state against the Mi'kmaq people seeking to exercise what they understood to be their court-affirmed treaty rights. Episodic violence continued through 2002.

The DFO, without community consultation, and without truly grappling with what treaty rights and Indigenous sovereignty might mean for the lobster fishery, shoehorned expanded Indigenous rights within

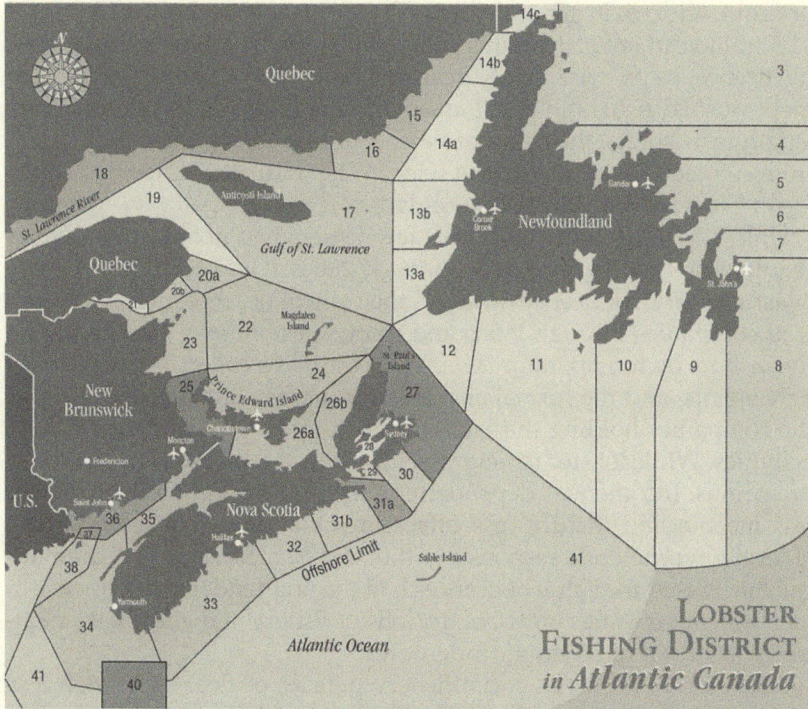

Figure 21. Map of lobster fishing districts in 2024 (Lobster Council of Canada). The boundaries still carry remnants of decisions made at the end of the nineteenth century.

Lobster Fishing Districts in 2024 (Lobster Council of Canada)

the existing framework of limited access through a property rights regime. Following *Marshall*, the federal government spent $545 million in the Maritimes for training programs and transferring licences to First Nation communities.[34] The influx of federal dollars to buy out non-Indigenous commercial fishermen and purchase Indigenous community commercial licences put further upward pressure on the price of lobster licences. This together with the increased value of lobsters led to a dramatic, almost inconceivable, upturn in the cost of licences in some Lobster Fishing Areas. Within a little more than a generation, lobster licences increased in value from a couple hundred dollars to a couple thousand dollars to a couple hundred thousand dollars to a peak of a million in LFA 34 and 35 in 2018.[35] A great deal of money is at stake for individual fishing enterprises with no guarantee of present or future returns. In general, it would require between $1.2 million and $1.5 million to enter the fishery depending on the boat, but including the licence, traps, rope, safety gear, and bait. This extraordinary investment pushed fishers in some areas into bigger boats. While there are length restrictions, the beam of vessels in LFA 34 has increased in size from eighteen to twenty-six feet wide.[36] The traditional Cape Sable boat looks like it has been put on steroids.

Unless a licence was inherited, few fishers had access to the credit required to enter the fishery. By the early 2000s, it was estimated that the "vast majority" of licences were in the control of large, highly capitalized companies through loans and mortgages, thereby circumventing the inshore owner-operator laws.[37] This created a covert system of trust agreements, and the formal and informal agreements usually granted the companies holding the loans decision-making power over fishing activities. While lobster fishing was lucrative, it remained economically precarious. Income was dependent on the catch; poor landings, meant less income and in turn less offseason Employment Insurance. The federal government's response to the *Marshall* decision that dramatically increased the value of licences had the unintended consequence of increasing corporate control of the lobster fishery and effectively creating a new form of economic indenture.[38]

But the DFO, with its social fishery policies of fleet separation and owner-operator requirements demanded that all controlling agreements be registered in 2007, and they were supposed to be phased out within seven years. Illegal secret arrangements continued, and in 2019 legislative changes enshrined fleet separation and the inshore owner-operator rules as part of the revised Fishery Act. But again, the central policy of the Atlantic inshore fishery was undermined by programs to support Indigenous lobster fishing. Indigenous Commercial Licences were exempt from both policies.[39]

Unlike individual lobster licences, communal licences held by First Nations can be leased and are tendered annually. These licences have become a vital income stream for many First Nations who face persistent poverty and lack resources to meet extensive community needs. In 2020 the annual lease for a licence in LFA 35 was $250,000 and $70,000 in LFA 34.[40] Few individuals have that kind of money to pay up front, so it is assumed that many of these licences are leased to corporations otherwise barred from the inshore fishery. In 2020 Wasoqopa'q (Acadia) First Nation, located in western Nova Scotia, took another approach for the eighteen Communal Commercial Licences issued within its territory in LFA 33 and 34. This First Nation leased them to individual band members for around $80,000–90,000 per season. How individuals in turn decided to fish the licences varied. The majority were re-leased for a profit to non-Indigenous commercial fishermen or fishing companies, but it was estimated that six captains and twenty-five to thirty deckhands who were band members found direct employment.[41]

Heinous expressions of racism between settler commercial fishermen and First Nations people occurred in the violence of both 1999–2000 and 2020. But to see only racism is to overlook the continued history of direct action associated with the lobster fishery. Violence and direct action re-emerge at moments when prices or catches are low and fishermen feel vulnerable. The collapse of prices in the spring of 2013 when the price dropped $3 to $4.25 a pound witnessed both a strike by Prince Edward Island lobstermen who refused to fish until prices increased and vigilante actions. Conflict on Nova Scotia's Eastern Shore during the 2015 season saw a flurry of trap cutting like that around Saddle Island in the 1930s.[42] Lobster fishermen continue to take the law into their own hands when the state is perceived as weak or out of touch with local concerns and context.

The events of the fall of 2020 did not erupt suddenly; rather they were deeply rooted in longstanding traditions and histories that shaped the actions and perceptions of both Indigenous communities and commercial fishermen. The issues at stake on both sides are far more complicated than what has generally been reported, and both sides shared a language of conservation and rights – albeit with very different meanings.[43] It should also go without saying, as noted at the time by Wasoqopa'q First Nation Deputy Chief Jeff Purdy, that there were individual bad actors on both sides.[44] In the frustration and destruction of property, it is difficult not to see the echoes of past descriptions of a three-corner fight between the state, fishermen, and packers, with this time Indigenous and commercial fishermen being set off against each other by the federal government. Colin Sproul – former president of the Bay of Fundy Inshore Fishermen's Association – condemned the federal government for not having "the political courage" to act. "The buck has stopped in these communities, and it's been left for us to sort out ... It's completely, patently unfair for the federal government to do that. They have to figure something out in Ottawa, and not leave it to us."[45] Echoes of past pleas to the federal government are also heard here.

Sipekne'katik First Nation symbolically launched its own managed moderate livelihood fishery on the twenty-first anniversary of the *Marshall* decision, and the first lobster tag was issued to Donald Marshall Jr.'s son.[46] It was operated out of Saulnierville on St. Mary's Bay where much of its FSC fishery took place and nearly a three-hour drive west of

Sipekne'katik. The FSC fishery, conducted by individuals at the household level, had been suspected as taking place on a commercial basis. In October 2017 DFO officers planted microchips in seventy-three lobsters located in the FSC traps of members of Sipekne'katik First Nation and traced the lobsters to the Halifax airport where they were about to be shipped to China. Sheng Ren Zheng and his company Guang Da International Limited were found guilty of illegal lobster buying in August 2020. As sales were illegal, the lobsters were sold at a discount. Reportedly, the price was only $3 a pound, less than half the commercial rate and potentially depressing all lobster prices.[47] There were rumours of a large-scale black market in operation that some Mi'kmaq fishermen understood as an expression of their sovereignty over resources. But extensive fishing out of season was seen by most non-Indigenous fishermen who held commercial licences to harm their property rights through undermining the local stock, undercutting prices, and damaging the reputation for a quality product.

Both sides are constrained by resource-extraction capitalism, but the Indigenous communities also carry the heavyweight of colonialism. This has led some First Nations to have little choice but to address immediate needs by accepting confidential Rights Reconciliations Agreements, time specific agreements that require Indigenous fishers to adhere to DFO regulations.[48] Other First Nations have developed their own moderate livelihood fishery conventions. There is no question about treaty rights under the Supreme Court's *Marshall* decision, but twenty-five years later there is still tension between *Marshall I* and its amendment *Marshall II*. There is no clear answer of what a moderate livelihood entails, if that restriction was appropriate, and why Indigenous communal licences would have been exempted from the foundational policies of the modern Atlantic inshore fishery.

Commercial fishermen, whom over the past fifty years had come to see lobster licences transformed into the most valuable personal property they were ever likely to own, were left as individuals holding the weight of Canadian colonialism. While the relatively small number of Indigenous traps being fished would not have an impact on the overall lobster fishery, this fishery is a profoundly local one and inshore lobsters do not migrate any great distance. Intense local fishing might diminish the stock in a particular place even if the impact on the entire fishery is negligible. The issue is that commercial lobstermen fish particular places, harbours, bays, sounds, shoals, and straits. As the economists of the 1950s and 1960s constantly argued, the intensification of fishing only makes the pieces of the cake smaller. Lobster fishing seasons emerged from a variety of social, biological, economic, and political factors, but claims about seasons not having *any* ecological conservation basis are dubious. Yes, this is about the quality of a hard shellfish and reproductive cycles, but it is also about catch rates, as lobsters seem to be hungry and more likely to be attracted to bait after they have moulted.[49]

For commercial settler fishermen traditional informal practices developed in conjunction with the formal regulations of a commercial fishery. Some informal practices were place- and community-specific, such as understandings of access rights and who could be kept out. After limited entry was introduced in the late 1960s, lobster was transformed from a poor people's fishery into a highly capitalized, lucrative enterprise, accessible to a limited number of community members. As it became more valuable, it was more closely regulated by the federal government and less closely by localized norms of conduct.[50] But not entirely, as the violence of the fall of 2020 indicated. With so much money now at stake, the lobster industry, much like the nineteenth century, has once again attracted large international corporations, now joined by new players with ties to organized crime.

How to keep corporations out of the inshore fishery will become more complicated with the sale of Clearwater Foods to Premium Brands Holding Corporation and a coalition of First Nations led by Membertou and Miawputek, including Sipekne'katik, Paqtnkek, Puksaqte'kne'katik, Potlotek, and We'koqma'q. Membertou had already leased private monopoly fish licences in LFA 41 offshore – a fishery that has always been controversial for commercial inshore fishermen and one that has been subject to serious violations of conservation practices.

The purchase of Clearwater also gestures towards the diversity in strategies to uphold treaty rights and sovereignty between different communities and divergent approaches to neoliberalism. Not all First Nations located in Mi'kma'ki joined this venture, just as not all Bands had signed agreements for communal commercial licences.[51] Arthur Bull, an advisor to the World Forum on Fisher Peoples described the issue as a tangled mess of treaty rights, human rights, and property rights.[52] Solving the problem is going to take political leadership, nation-to-nation negotiations, and close attention to local contexts. The problem is not *Marshall* – rather the federal government's response to *Marshall* that attempted to avoid big issues of Indigenous sovereignty and shoehorn a response into existing categories of conceiving licences as property. This was completely consistent with other contradictions in lobster policy – promoting a privatized, market-regulated fishery with social fishery, welfare objectives.

From the moment the commercial lobster fishery took off in the 1870s, the new Canadian state attempted to manage it, trying to balance its political interests with the interests of transnational and local packers, fishermen, and lobsters to maximize economic and partisan profits while maintaining resource sustainability. The complex regulatory framework that emerged reveals both the powers and limits of the state's ability to know, to expand, and to employ coercive power, as regulations often lack legitimacy and were ignored. State formation at the local level could be precarious. As a dispersed and decentralized resource, lobster may have been especially resistant to centralized expert management and local knowledge continues to be relevant in making good policy decisions. Lobster provides an interesting case in exploring how the distant Canadian state sought to "know" the resource – and how knowledge gathering served political agendas beyond identifying the best policy solutions.

Local communities responded to regulations with a complicated mix of formal and informal practices that reflected their specific context and interests. Laws were often rejected, and local versions of justice were pursued through violent direct action or arson. The particularities of local understandings of rights and usage meant that while illegal fishing remained widespread, locally held notions of the commons or access rights could lead communities to request state interventions to uphold what they regarded as their legitimate, if unofficial, rights. Over time the focus of state regulations shifted from regulating packers to policing fishermen, and the coercion associated with destruction or confiscation of property, fines, or jail moved to education, with an emphasis on lobstermen's individual self-interest in preserving the stock.

Before the First World War, lobster politics played a significant electoral role in the formation of federal governments. As the Atlantic region's political importance declined, lobster politics shifted towards cooperatives and associations – organizations increasingly entwined with the state. After the Second World War, the federal government's interest in lobster policy also had social welfare dimensions beyond the ad hoc use of regulatory exceptions to provide relief in times of need. The focus shifted from stock conservation to the living standards of fishermen and their families. Economists who gained influence within the Department of Fisheries focused on efficiency – understood as fewer fishermen catching more lobsters. The focus on regional development, income improvement, and specialization reframed community-based access rights as personal property. Lobster became a harbinger of broader changes in ocean governance. Policy agendas that initially emphasized social welfare opened the door to a new neoliberal common sense – one that recast access through the lens of market-based individual property rights.

On the last Monday of November in 2020, I was at the wharf before 7:00 a.m. along with many family and community members to wish the local LFA 33 lobster fishermen a safe and prosperous season as they set out on what is called locally dumping day. It had been a difficult and tense fall, with conflict and violence in neighbouring LFA 34 – the wealthiest fishing area – between settler commercial fishermen and Indigenous fishermen from Sipekne'katik pursuing their autonomous moderate "living" lobster fishery. As the boats lined up in position to start for their grounds, older men on the wharf pointed to the various boats offering the pedigree in terms of licence ownership. That boat is fishing my father's licence or that boat used to be so-and-so's licence. Here the licences were being discussed almost exactly in the same terms as local people would discuss a piece of land or house. The ownership of a lobster licence was real and seemingly permanent property – not at all like an annual hunting licence that was renewed with a fee, or a licence connected to another fishery that came with a quota announced annually. No one there would see it as a privilege. It was a right awarded by inheritance, hard work, or borrowing.

Dumping day on the same wharf three years later brought great change, as retirements and death meant that fewer of the boats were operated by men who lived in the local community. This was partly generational change and not to be extracted on a larger scale, but it is striking how many of those leaving the wharf that morning had arrived the week before from Prince Edward Island, a good four hours' drive away. In those three years, change and circumstances removed opportunities for local succession to younger people who wanted to fish.

The excitement of dumping day and the hopes it carries continues to make me wonder – as I have wondered over the last fifteen years. Do I have the privilege of watching the last days of another dying commercial fishery? Some of the anxieties and uncertainties are specific to current world events. Would the cost of diesel force lobster prices too high? Would Chinese markets continue to expand? Would the United States use Canadian lobster as a political tool in a populist trade war? Would expanded large-scale, corporate open-pen, fin-fish aquaculture destroy prime fishing areas? Other concerns are more daunting, especially those connected to climate change, the availability of bait, and North Atlantic currents and acidification.[53] The water temperatures in the Gulf of Maine are warming faster than almost any ocean waters in the world, and lobster stocks are plummeting in southern New England.

In this context, Canadian political uncertainties such as failing policies aimed at protecting the independence of the inshore fishery and keeping the lobster fishery out of corporate hands (and with it, the siphoning of revenues away from local communities) – remain pressing concerns. Equally unresolved is the persistent lack of political leadership in addressing and implementing fair, sustainable approaches to the Indigenous fishery. While the "lobster question" has evolved over the last 150 years, it continues to be shaped as much by local politics as economics or environmental considerations.

Notes

Introduction

1 In addition, there are sixteen underwater topographies with bars, gullies, holes, ledges, rocks, and shoal with lobster in their names. Canadian Geographic Names Database, www.nrcan.gc.ca/earth-sciences/geography /querying-canadian-geographical-names-database/canadian-geographical -names-database/19870.

2 Stewart Bates, *Report on the Canadian Atlantic Sea-Fishery: Royal Commission on Provincial Development and Rehabilitation*, IX (Halifax: King's Printer, 1944), 10.

3 In 1939 fishermen and fish handlers in Lockeport, Nova Scotia, attempted to have the Canadian Fishermen's Union recognized as their official bargaining agent. Sue Calhoun, *The Lockeport Lockout: An Untold Story in Nova Scotia's Labour History* (Halifax: Lockeport Lockout, 1983). The history of fishermen, unions, and politics on the Pacific Coast is completely different. See Homer Stevens and Rolf Knight, *Homer Stevens: A Life in Fishing* (Madeira Park, BC: Harbour Publishing, 1992). For Newfoundland, see Parzival Copes, "The Fishermen's Vote in Newfoundland," *Canadian Journal of Political Science* 3, no. 4 (December 1970): 579–604; Melvin Baker, "Challenging the 'Merchants' Domain': William Coaker and the Price of Fish, 1908–1919," *Newfoundland & Labrador Studies* 29, no. 2 (Fall 2014): 189–226; David Alexander, "The Political Economy of Fishing in Newfoundland," *Journal of Canadian Studies* 11, no. 1 (April 1976): 32–40. More recent work by Daniel Banoub focuses on ecological politics; see his "Black Monday, 1894: Saltfish, Credit, and the Ecology of Politics in Newfoundland," *Atlantic Studies* 17, no. 2 (June 2020): 227–43.

4 Joseph Gough, *Managing Canada's Fisheries: From Early Days to the Year 2000* (Quebec City: Septentrion, 2006), 124.

5 Former territories that became Alberta, Manitoba, and Saskatchewan
waited until 1930 to gain control of their natural resources through the
Natural Resources Transfer Act (1930).

6 J.H. Chisholm, ed., *The Speeches and Public Letters of Joseph Howe* (Halifax:
Chronicle Publishing Company, 1909), II, 515.

7 See, for example, Martha Walls, *No Need of a Chief for This Band: The
Maritime Mi'kmaq and Federal Electoral Legislation, 1899–1951* (Vancouver:
UBC Press, 2010); William C. Wicken, *The Colonization of Mi'kmaw Memory
and History, 1794–1929: The King v. Gabriel Sylliboy* (Toronto: University
of Toronto Press, 2012); Bill Parenteau, "Care, Control, and Supervision:
Native People in the Canadian Atlantic Salmon Fishery, 1867–1900,"
Canadian Historical Review 79, no. 1 (March 1998): 1–35; Bill Parenteau
and James Kenny, "Survival, Resistance, and the Canadian State: The
Transformation of New Brunswick Native Economy, 1867–1930," *Journal
of the Canadian Historical Association* 13, no. 1 (2002): 9–71; and Andrew
Parnaby, "The Cultural Economy of Survival: The Mi'kmaq of Cape Breton
in the Mid-19th Century," *Labour/Le Travail* 61, (2008): 69–98.

8 There is a very large literature on how state and communities are *both*
sources of property and usage rights. The work of James Acheson on the
Maine lobster fishery and Ralph Matthew's work on the Newfoundland
fishery is especially important. Anthropologist Anthony Davis, working
on the restricted access post-1970 period, argues that informal individual
rights emerged from community rights that emphasized economic
dependence, proximity to the fishing port, and continuous occupation
and use. James M. Acheson, "The Lobster Fiefs: Economic and Ecological
Effects of Territoriality in the Maine Lobster Industry," *Human Ecology*
3 (July 1975): 183–207; Ralph Matthews, *Controlling Common Property:
Regulating Canada's East Coast Fishery* (Toronto: University of Toronto
Press, 1993); Anthony Davis, "Property Rights and Access Management
in the Small Boat Fishery: A Case Study from Southwest Nova Scotia,"
in *Atlantic Fisheries and Coastal Communities: Fisheries Decision-Making
Case Studies*, ed. Cynthia Lamson and Arthur J. Hanson (Halifax: Ocean
Studies Programme, Dalhousie University, 1984), 133–64; Anthony Davis
and Leonard Kasdan, "Bankrupt Government Policies and Belligerent
Fishermen's Responses: Dependency and Conflict in the Southwest Nova
Scotia Small Boat Fisheries," *Journal of Canadian Studies* 19, no. 1 (February
1984): 108–24.

9 Canada, Lobster and Oyster Commission, *Report of the Commissioners
Appointed by His Excellency the Governor General in Council, of Date 4th July,
1887, to Enquire into and Report upon the Lobster and Oyster Fisheries of Canada*
(Ottawa, 1887), 14; See also Suzanne Morton, "'Alien Concerns': American
Canners in the Gulf of St Lawrence Lobster Fishery, 1870–1914," in *The*

Greater Gulf: Environmental History and the Gulf of St Lawrence, ed. Claire Campbell, Edward MacDonald, and Brian Payne (Montreal and Kingston: McGill-Queen's University Press, 2020), 224–59.

10 Garrett Hardin, "The Tragedy of the Commons," *Science* 162, no. 3859 (December 1968): 1243–8. Elinor Ostrom's work is the most important to challenge this view and argues that communities could and did collectively manage resources. Ostrom's scholarship included work on the Maine lobster fishery; see Edella Schlager and Elinor Ostrom, "Property-Rights and Natural Resources: A Conceptual Analysis," *Land Economics* 68, no. 3 (August 1992): 249–62. Paul Dragos Aligica and Ion Sterpan, "Governing the Fisheries: Insights from Elinor Ostrom's Work," in *Institutions and Policies*, ed. R. Wellings (London: Institute of Economic Affairs, 2017), 96–117.

11 Edward E. Prince to T.B. Flint, 27 July 1896, RG 23, Fisheries & Oceans (FO), vol. 160, T-2848, File 508, Parts: 1 Changes to Lobster Season – Nova Scotia 1894–1897, Library Archives Canada (LAC), Ottawa.

12 J.T. Croteau, *Cradled in the Waves: The Story of a People's Co-operative Achievement in Economic Betterment on Prince Edward Island*, Canada (Toronto: Ryerson Press, 1951), 5, 84–5.

13 Croteau, *Cradled in the Waves*, 85; Canada, *Commission to Inquire into the Herring and Sardine Industry of the Bay of Fundy, as Well as into the Ravages of the Dog-Fish and the General Condition of the Lobster Fishery at the Magdalen Islands, St. Mary's Bay and The Bay of Fundy* (1905), 5–6, accessed 18 July 2020, https://epe.lac-bac.gc.ca/100/200/301/pco-bcp/commissions-ef/tucker1905-eng/tucker1905-eng.html.

14 This crucial point was also made by John Wagner and Anthony Davis, "Property as a Social Relation: Rights of 'Kindness' and the Social Organization of Lobster Fishing among Northeastern Nova Scotian Scottish Gaels," *Human Organization* 63, no. 3 (Fall 2004): 320–33.

15 E.P. Thompson, "Custom Law and Common Right," *Customs in Common* (Harmondsworth: Penguin 1991), 97.

16 Brian Payne, "Local Economic Stewards: The Historiography of the Fishermen's Role in Resource Conservation," *Environmental History* 18, no. 1 (January 2013): 35.

17 Elinor Ostrom, *Governing the Commons: The Evolution of Institutions for Collective Action* (Cambridge: Cambridge University Press, 1990).

18 Kurt Korneski, "Development and Diplomacy: The Lobster Controversy on Newfoundland's French Shore, 1890–1904," *International History Review* 36, no. 1 (2014): 45–69; Department of Fisheries and Oceans (DFO), Newfoundland and Labrador Region, "An Assessment of American Lobster in Newfoundland," Canadian Science Advisory Secretariat Science Advisory Report 2006/009, 3–4.

19 Brian Payne, *Eating the Ocean: Seafood and Consumer Culture in Canada* (Montreal and Kingston: McGill-Queen's University Press, 2022).

20 See Richard W. Judd, "Saving the Fisherman as Well as the Fish: Conservation and Commercial Rivalry in Maine's Lobster Industry, 1872–1933," *Business History Review* 62, no. 4 (Winter 1988): 596–625.

21 "Decline of the Canadian Lobster Trade," *Sun* [Baltimore, MD], 10 October 1883, p. 2.

22 See chapter 6.

23 See Trevor A. Branch and Danika Kleiber, "Should We Call Them Fishers or Fishermen," *Fish and Fisheries* 18, no. 1 (January 2017): 114–27; Ilima Loomis, "Fishers or Fishermen – Which Is Right?" *Hakai* 13 October 2015, accessed 24 July 2022, https://hakaimagazine.com/news/fishers-or-fishermen-which-right/; Sarah Sweet, "Why Women Who Fish Are Still Fishermen," *Walrus*, Updated 20 November 2021, Published 27 April 2017. Accessed 24 July 2022, https://thewalrus.ca/why-women-who-fish-are-still-fishermen/. There is general agreement that the term fishermen continues to be the preferred term by both men and women in the industry, and the Maritime Fishermen's Union has not changed its name. The Federal Government started using "Fish Harvester" in the early 2000s.

24 Diary of William K. McClearn, Little Harbour, Shelburne County, transcript 1999, B1844, Shelburne County Genealogical Society, Shelburne, NS.

25 For example, see "Wage Summary 1886," Series A, General Store & Warehouse/Factory, Subseries 8 Ledger 1886–7, MG200 Henry T. d'Entremont Fonds, Argyle Township Courthouse & Archives, Tusket, NS; Chatham Individual Payroll Cards, 1939–56, W.S. Loggie Co. Ltd. Fonds MC1049 MS1 I, Provincial Archives of New Brunswick, Fredericton, NB.

26 "Who's Who in the Fishing World," *Canadian Fisherman* (*CF*), July 1915, p. 208.

27 There is no broad comprehensive history of the Canadian lobster fishery and so understandings of resource regulation come from work such as the following: Sean Cadigan, "The Moral Economy of the Commons: Ecology and Equity in the Newfoundland Cod Fishery, 1815–1855," *Labour/Le Travail* 43 (Spring 1999): 9–42; Ian J. Jesse, "A 'Game War' in the Borderlands: Cross-Border Poaching in the Northeast 1886–1908," *American Review of Canadian Studies* 48, no. (June 2018): 152–62; Richard Judd, *Common Lands, Common People: The Origins of Conservation in New England* (Cambridge: Harvard University Press, 1997); Richard W. Judd, "Grassroots Conservation in Eastern Coastal Maine: Monopoly and the Moral Economy of Weir Fishing, 1893–1911," *Environmental Review* 12 (Summer 1988): 80–103; Tina Loo, *States of Nature: Conserving Canada's Wildlife in the Twentieth Century* (Vancouver: UBC Press, 2007); Bill Parenteau, "A Very Determined Opposition to the Law: Conservation, Angling Leases, and Social Conflict in the Canadian Atlantic Salmon

Fishery, 1867–1914," *Environmental History* 9, no. 3 (July 2004): 436–63. The work of Harold A. Innis remains foundational – *The Fur Trade in Canada: An Introduction to Canadian Economic History* (Toronto: University of Toronto Press, 1940, rev. ed. 1956) and *The Cod Fisheries: The History of an International Economy*, revised ed. (Toronto: University of Toronto Press, 1954) – and there is a growing literature on Indigenous resources to state resource regulations; see Shelley Denny and Lucia Fanning, "Balancing Community Autonomy with Collective Identity: Mi'kmaq Decision-Making in Nova Scotia," *Canadian Journal of Native Studies* 36, no. 2 (2016): 81–106 and Howard Ramos and Caitlin Krause. "Sharing the Same Waters," *British Journal of Canadian Studies* 28, no. 1 (May 2015): 23–4. Although not an overview of the lobster industry, Kurt Korneski's "Development and Degradation: The Emergence and Collapse of the Lobster Fishery on Newfoundland's West Coast, 1856–1924," *Acadiensis* 41, no. 1 (Winter/Spring 2012): 21–48 merits particular mention.

28 James M. Acheson, *The Lobster Gangs of Maine* (Lebanon, NH: University Press of New England, 1988); *Capturing the Commons: Devising Institutions to Manage the Maine Lobster Industry* (Lebanon, NH: University Press of New England, 2003); and Roy J. Gardner, "Strategies, Conflict and the Emergence of Territoriality: The Case of the Maine Lobster Industry," *American Anthropologist* 106, no. 2 (June 2004): 296–307; Jennifer F. Brewer, "Don't Fence Me In: Boundaries, Policy, and Deliberation in Maine's Lobster Commons," *Annals of the Association of American Geographers* 102, no. 2 (March 2012): 383–402; Judd, "Saving the Fisherman" and Colin Woodward, *The Lobster Coast: Rebels, Rusticators, and the Struggle for a Forgotten Frontier* (New York: Penguin Books, 2004).

29 Among the best local studies is Donald W. Jacquard, *Lobstering, Southwestern Nova Scotia, 1848–2009* (Lower Wedgeport, NS: Donald W. Jacquard, 2009). There is parallel work done on Prince Edward Island by Sharon Arsenault: "Packing Lobsters at the Beach: Cannery Life on Prince Edward Island," *Island Magazine*, no. 60 (September 2006): 13–21.

30 See Régis Brun, *La ruée vers le homard des Maritimes* (Moncton: Michel Henry, 1988); "L'industrie du homard dans le Sud-Est acadien du Nouveau-Brunswick, 1850–1900," *Égalité* 16 (Fall 1985): 17–33; Nicolas Landry, *Les pêches dans la péninisule acadienne, 1850–1900* (Moncton: Éditions d'Acadie, 1994).

31 For histories of the fisheries in Canada, see Richard Apostle and Gene Barrett, eds., *Emptying Their Nets: Small Capital and Rural Industrialization in the Nova Scotia Fishing Industry* (Toronto: University of Toronto Press, 1992); Daniel Banoub, *Fishing Measures: A Critique of Desk-Bound Reason* (St. John's: Memorial University Press, 2021); Dean Bavington, *Managed Annihilation: An Unnatural History of the Newfoundland Cod Collapse*

(Vancouver: UBC Press, 2010); Sean Cadigan, *Hope and Deception in Conception Bay: Merchant-Settler Relations in Newfoundland, 1785–1855* (Toronto: University of Toronto Press, 1995); James Candow and Carol Corbin, eds., *How Deep Is the Ocean? Essays on Canada's Atlantic Fishery* (Sydney, NS: University College of Cape Breton Press, 1997); Matthew Evenden, *Fish Versus Power: An Environmental History of the Fraser River* (Cambridge: Cambridge University Press, 2004); Michael Harris, *Lament for an Ocean: The Collapse of the Atlantic Cod Fishery* (Toronto: McClelland and Stewart, 1998); Dianne Newell, *Tangled Webs of History: Indians and the Law in Canada's Pacific Coast Fisheries* (Toronto: University of Toronto Press, 1993); Dianne Newell and Rosemary Ommer, eds., *Fishing Places, Fishing People: Traditions and Issues in Canadian Small-Scale Fisheries* (Toronto: University of Toronto Press, 1999); Alex Rose, *Who Killed the Grand Banks?: The Untold Story behind the Decimation of one of the World's Greatest Natural Resources* (Mississauga: John Wiley and Sons, 2008); Lissa K. Wadewitz, *The Nature of Borders: Salmon, Boundaries, and Bandits on the Salish Sea* (Vancouver: UBC Press, 2012), and Miriam Wright, *A Fishery for Modern Times: The State and the Industrialization of the Newfoundland Fishery, 1934–1968* (Toronto: Oxford University Press, 2001). A recent rich history of lobster and Indigenous issues is developing in Sarah J. King, *Fishing in Contested Waters: Place and Community in Burnt Church/Esgenoôpetitj* (Toronto: University of Toronto Press, 2014).

32 Gough, *Managing Canada's Fisheries*; Douglas Harris, *Fish, Law and Colonialism: The Legal Capture of Salmon in British Columbia* (Toronto: University of Toronto Press, 2001); Jennifer Hubbard, *A Science on the Scales: The Rise of Canadian Atlantic Fisheries Biology, 1898–1939* (Toronto: University of Toronto Press, 2006); Kenneth Johnstone, *The Aquatic Explorers: A History of the Fisheries* (Toronto: University of Toronto Press and Research Board of Canada, 1977); L.S. Parsons, *Management of Marine Fisheries in Canada* (Ottawa: National Research Council of Canada and DFO, 1993).

33 H.V. Nelles, *The Politics of Development: Forests, Mines, and Hydro-Electric Power in Ontario, 1849–1941* (Toronto: Macmillan, 1974). Also, Joseph E. Taylor, *Making Salmon: An Environmental History of the Northwest Fisheries Crisis* (Seattle: University of Washington Press, 1999) for the American example. See also Shannon Stunden Bower, *Wet Prairie: People, Land, and Water in Agricultural Manitoba*, Nature, History, and Society Series (Vancouver: UBC Press, 2011). Robert D. Cairns, "Natural Resources and Canadian Federalism: Decentralization, Recurring Conflict, and Resolution," *Publius* 22, no. 1 (Winter 1992): 55–70 is firmly rooted in traditional political economy.

34 Stéphane Castonguay, *Le gouvernement des ressources naturelles: sciences et territorialités de l'État québécois, 1867–1939* (Quebec City: Presses de l'Université Laval, 2016) translated and published as *The Government of*

Natural Resources: Science, Territory, and State Power in Quebec, 1867–1939, translated by Roth Käthe (Vancouver: UBC Press, 2021).

35 Loo, *States of Nature*.

36 See, for example, John Sandlos and Arn Keeling, eds. *Mining and Communities in Northern Canada: History, Politics and Memory* (Calgary: University of Calgary Press, 2015), Caroline Desbiens, *Power from the North: Territory, Identity, and the Culture of Hydroelectricity in Quebec* (Vancouver: UBC Press, 2013) and Brittany Luby, *Dammed: The Politics of Loss and Survival in Anishinaabe Territory* (Winnipeg: University of Manitoba Press, 2020).

37 A superb exploration of the centrality of the "local" and the specific in Indigenous history is provided by Mercedes Peters, "The Future is Mi'kmaq: Exploring the Merits of Nation-Based Histories as the Future of Indigenous History in Canada," *Acadiensis* 48, no. 2 (Autumn 2019): 206–16.

38 R.W. Sandwell, *Canada's Rural Majority: Households, Environments, and Economies, 1870–1940* (Toronto: University of Toronto Press, 2016), 5.

39 James Murton, *Canadians and Their Natural Environment: A History* (Don Mills, ON: Oxford University Press, 2021), 112.

40 Fred Burrill, Re-developing Underdevelopment: An Agenda for New Histories of Capitalism in the Maritimes *Acadiensis* 48, no. 2 (Autumn 2019): 180.

41 For forestry see Mark J. McLaughlin and Bill Parenteau, "A 'Fundamental Cost that We Can't Deal With'?: The Political Economy of the Pulp and Paper Industry in New Brunswick, 1960-Present," in *Exploring the Dimensions for Self-Sufficiency in New Brunswick*, ed. Michael Boudreau, Peter G. Toner, and Tony Tremblay (Fredericton: New Brunswick and Atlantic Canada Studies Research and Development Centre, 2009), 13–34; Mark McLaughlin "'Trees are a Crop': Crown Lands, Labour, and the Environment in New Brunswick's Forest Industries, 1940–1982." (PhD diss., University of New Brunswick, 2014); William M. Parenteau, "Forest and Society in New Brunswick: The Political Economy of the Forest Industries, 1918–1939" (PhD diss., University of New Brunswick, 1994); Bill Parenteau and L. Anders Sandberg, "Conservation and the Gospel of Economic Nationalism: The Canadian Pulpwood Question in Nova Scotia and New Brunswick, 1919–1933," *Environmental History Review* 19, no. 2 (Summer 1995): 55–83; Bill Parenteau, "Looking Backward, Looking Ahead: History and Future of the New Brunswick Forest Industries" *Acadiensis* 42, no. 2 (September 2013), 92–113; L. Anders Sandberg, ed., *Trouble in the Woods: Forest Policy and Social Conflict in Nova Scotia and New Brunswick* (Fredericton: Acadiensis Press, 1992) and L. Anders Sandberg and Peter Clancy, *Against the Grain: Foresters and Politics in Nova Scotia* (Vancouver: UBC Press, 2000)

42 Burrill, "Re-developing Underdevelopment," 188. This paragraph owes much to my colleague Don Nerbas.

Chapter One

1 "Vibrio," accessed 11 May 2022, www.canada.ca/en/public-health/services /food-poisoning/vibrio.html.

2 Hubert J. Squires, "*Decapod Crustacea* of the Atlantic Coast," *Canadian Bulletin of Fisheries and Aquatic Sciences* 221 (Ottawa: DFO, 1990): 325–32.

3 "Lobster Research at the University of New Hampshire," Frequently Asked Questions, accessed 21 May 2022, www.lobsters.unh.edu/offshore _fishery/faq/faq.html; "Fun Facts about Lobster," NOAA Fisheries, accessed 21 May 2022, www.fisheries.noaa.gov/national/outreach-and-education/fun-facts-about-luscious-lobsters#how-far-do-lobsters-travel? H.O. Haakonsen and A.O. Anoruo, "Tagging and Migration of the American Lobster *Homarus americanus*," *Reviews in Fisheries Science* 2, no. 1 (1994): 79–93.

4 John Mark Hanson, "Predator-Prey Interactions of American Lobster *(Homarus americanus)* in the Southern Gulf of St. Lawrence, Canada," *New Zealand Journal of Marine and Freshwater Research* 43, no. 1 (2009): 69–88; Erin B. Wilkinson, Jonathan H. Grabowski, Graham D. Sherwood, and Philip O. Yund, "Influence of Predator Identity on the Strength of Predator Avoidance Responses in Lobsters," *Journal of Experimental Marine Biology and Ecology* 465 (2015): 107–12; Stephanie Boudreau and Boris Worm, "Top-Down Control of Lobster in the Gulf of Maine: Insights from Local Ecological Knowledge and Research Surveys," *Marine Ecology Progress Series* 403 (2010): 181–91.

5 "Assessment of Lobster (*Homarus americanus*) in Lobster Fishing Area 34," DFO Canadian Science Advisory Secretariat, Science Advisory Report 2021/015, April 2021; Trevor Corson, *The Secret Life of Lobsters: How Fishermen and Scientists are Unravelling the Mysteries of our Favourite Crustacean* (New York: HarperCollins, 2004); Nancy Frazier, *I, Lobster: A Crustacean Odyssey* (Durham: University of New Hampshire Press, 2012); Elizabeth Townsend, *Lobster: A Global History* (London: Reaktion Books, 2011). See also Gwynn Guilford, "The Enigma Behind America's Freak 20-Year Lobster Boom," *Quartz*, 6 October 2015, accessed 12 March 2022, https://qz.com/506376/lobsters-2/.

6 "The American Lobster," accessed 12 March 2022, www.parl.ns.ca/lobster /overview.htm; Rhode Island Sea Grant, "The American Lobster," accessed 12 May 2022, https://web.archive.org/web/20110716234846/http:// seagrant.gso.uri.edu/factsheets/fslobster.html. "Life of the American Lobster – Life Cycle & Reproduction," Lobster Institute, University of Maine, accessed 12 March 2022, http://umaine.edu/lobsterinstitute /education/life-of-the-american-lobster/life-cycle-reproduction/.

7 Beothuk stored lobster tails and dried the meat. Shanawdithit's Sketch, "Different Kinds of Animal Foods," in John P. Howley, *The Beothuks or Red Indians: The Aboriginal Inhabitants of Newfoundland* (Cambridge: Cambridge

University Press, 1915), following 246; Ehud Spanier et al., "A Concise Review of Lobster Utilization by Worldwide Human Populations from Prehistory to the Modern Era," *ICES Journal of Marine Science* 72, no. 1 (July 2015): 15.

8 F.G. Speck and R.W. Dexter, "Utilization of Animals and Plants by the Micmac Indians of New Brunswick," *Journal of the Washington Academy of Sciences* 41, no. 8 (1951): 250–9; Natalie B. Stoddard, *Micmac Foods* (Halifax: Halifax Natural Science Museum, 1967?).

9 A.J.B. Johnston, "The Early Days of the Lobster Fishery in Atlantic Canada," *Material History Bulletin* 33 (Spring 1991): 56–60.

10 T.T. Pigot, "Mount Stewart," p. 33, 2009-H0014, #2 Food processing PEI, Peter Rider Fonds Research Material. Canadian Museum of History Archives (CMHA), Gatineau, QC.

11 Canada, Lobster and Oyster Commission, *Report*, 6.

12 Canada, Department of Marine & Fisheries (M&F), "Special Report on Certain Petitions Against the Regulation Affecting the Lobster Fishery," Appendix O, *Annual Report*, (Ottawa, 1873), 149.

13 Canada, *Annual Report of the Department of M&F for the Year Ending 30th June 1871* (Ottawa, 1871), 132.

14 W. Jeffrey Bolster, *The Mortal Sea: Fishing the Atlantic in the Age of Sail* (Cambridge, MA: Harvard University Press, 2014), 213.

15 W.F. Tidmarsh, Portland Packing Co, Charlottetown to Chairman, M&F Committee, 4 November 1909, MG6 A, vol. 24, 3, Measurement of Lobsters. Roberts, Simpson and Co of Liverpool, GB and Halifax, NS (RS&Co.), Nova Scotia Archives (NSA), Halifax, NS.

16 Massachusetts, Commissioners of Fisheries and Game, *The Lobster Fishery: A Special Report. Suggestions for Uniform Laws Made to the Legislature of Massachusetts by the Commissioners of Fisheries and Game 1911* (Boston: Wright and Potter Co, 1911), 12, 19.

17 *The Graphic* (London) 23, no. 590 (19 March 1881): 267.

18 Thomas F. Knight, *Shore and Deep Sea Fisheries of Nova Scotia* (Halifax, NS: A. Grant, 1867), 57–8.

19 Fish Canners Section, Draft 10th Article, H.E. Baker, "Early American Canners at Home and Abroad," 3, MG6A, vol. 24, 10 RS&Co., Canadian Lobster Fishery Statistics and other data from 1869–96, clippings, NSA.

20 William Wakeham memo, 15 March 1894, 5 February 1894, RG23, FO, vol. 181, T 2947, File 723, Parts 1, Trawling for Lobster, 1893–7, LAC.

21 Charles Stayner report, Halifax, NS, 19 September 1894; RG23, FO, vol. 253, T 3173, File 1645 Reports on Lobsters 1894; W.A. Found to Wakeham, 25 March 1912, RG2, FO, vol. 298, T 3981, File: 2314, Part 10A, Lobster Regulations – Minister's File, 1912, LAC.

22 From Wakeham, 1 April 1912; Found to Wakeham, 18 April 1912; RG23, FO, vol. 298, T 3981, File 2314, Part 11, Lobster Regulations 1912–14; Found

to R.A. Chapman, January 1909, RG23, FO, vol. 297, T 3981, File 2314, Part 9, Lobster Regulations 1909–10, LAC.

23 Canada, *Report of the Lobster Industry*, 1892 (Ottawa: S.E. Dawson, 1893), Appendix C, 22.

24 Canada, Commission on the Lobster Industry in Quebec and Maritime Provinces, *Report of Commander William Wakeham, Officer in Charge of the Gulf Fisheries Division, Province of Quebec, of an Investigation into the Lobster Fishery Pursuant to Order in Council dated June 21, 1909* (Ottawa: C.H. Parmalee, 1910), 5.

25 Canada, *Report of the Canadian Lobster Commission 1898* (Ottawa: S.E. Dawson, 1899), 26; John David Flint, "The Lobster Fishery of Southwest Nova Scotia: A Case Study of the Effects of Structural Transformation on the Allocation of Access to a Publicly Owned Resource," (PhD diss., Dalhousie University, 2002), 121; Census of Canada, New Brunswick, District 24, Westmorland, A, Botsford, 1891, p. 33. Edward Wheeler, 15, born in England, labourer working in household with fishermen.

26 "Parlour Lobster Traps," *Popular Mechanics*, September 1906, p. 920.

27 Canada, *Royal Commission on Price Spreads of Food Products*, Andrew Stewart, Chairman, vol. 2 (Ottawa, Queen's Printer, 1960), 255.

28 For example, see "Progress in Insurance Funds," *CF*, December 1954, 23; "Space-Saving Dome-Shaped Traps Arouse Lobster Fishermen's Interest," November 1964, 22; "Du Pont of Canada Invites You to Preview the New Lobster Traps of 'Sclair' Polyethylene," 15 September 1964; To E. Gosse, Deputy Minister (DM) Department of Fisheries 18 April 1963, GN34.2 18/21 Lobster Trap Designs and Designers, Provincial Archives of Newfoundland and Labrador (PANL), St. John's, NFL.

29 Wakeham, *Report*, 5.

30 Max Clarke, "Make and Break Engines," Memorial University Intangible Cultural Heritage, accessed 19 March 2022, www.mun.ca/ich/inventory/MaxClarkeProfile.php.

31 Arch J. Macdonald to Howard, 23 May 1908, Acc 4225/5/1/41, Macdonald family [Georgetown] fonds, Prince Edward Island Public Archives and Record Office (PARO), Charlottetown, PEI.

32 Jennifer F. Brewer, "Governing the Fishing Commons: Institutions, Ecosystems and Democracy in the Co-Management of Maine Lobster and Groundfish," (PhD diss., Clark University, 2007), 48.

33 Abram Hatfield to DM M&F, 28 October 1911, RG2, FO, vol. 17, T 2939, File 673, Part 2 1910–13, Charter of Steamers to Protect Maritimes' Lobster Fishery, 1894–1913, LAC.

34 Canada, Department of M&F, *Fisheries Annual Report 1907* (Ottawa: S.E. Dawson, 1907), 34.

35 Canada, Royal Commission Investigating the Fisheries of the Maritime Provinces and the Magdalen Islands, 1928 (A.K. Maclean Commissioner,

hereafter Maclean Commission) Evidence, Shelburne, NS, 9 December 1927, testimony of Arnold Newell, 3161 MG6, vol. 20, NSA.

36 "Four Admit to Washing Berried Lobsters," *Halifax Herald*, 10 May 1937, p. 3.

37 Brewer, "Governing the Fishing Commons," 50.

38 Beach Point, PEI, 29 August 1898, RG23, FO, vol. 284, T 3207, File 2154, Illegal Lobster Fishing, 1898, LAC.

39 *Report on Illegal Fishing and Canning of Lobsters and Illegal Fishing of Smelts in Lobster Fishing Districts Nos. 7 and 8* To Hon. J.E. Michaud, M.P., Minister of Fisheries, by Hon. Arthur T. LeBlanc, Commissioner (Ottawa: J.O. Patenaude, 1938).

40 RG23, FO, vol. 189, T 2955, File 830, Parts 1, Herrings and Sardine Industry in New Brunswick – Use of Herrings as Lobster Bait 1894–1912, LAC.

41 "Department Acts to Avert Bait Shortage Crisis," *Trade News* 1, 7 (January 1949), p. 1.

42 For the Pacific Ocean, see Benjamin Bryce, "Japanese Exclusion and Environmental Conservation in the BC Salmon Fisheries, 1900–1930," *Western Historical Quarterly* 53, no. 3, (Autumn 2022): 267–92.

43 African Nova Scotian fishermen in Shelburne County (Canada, 1881 Census, Nova Scotia, District 13 Shelburne, Subdistrict K, Port Latour, pp. 32–5) One lobster packer – 45-year-old Richard Hamilton of Port Latour, 1881, http://data2.collectionscanada.gc.ca/e/e325/e008119799.jpg.

44 S. Hollingworth, *The Present State of Nova Scotia: A Brief Account of Canada and the British Islands on the Coast of North America* (Edinburgh: William Creech, 1787), 63.

45 Canada, 1871 Census, New Brunswick, District 184 Northumberland, Subdistrict A Alnwick, p. 28. http://data2.collectionscanada.ca/1871/pdf/4396660_00408.pdf.

46 Canada, "Report of Indian Affairs," *Sessional Papers of the Dominion of Canada: Volume 4, First Session of the Fifth Parliament, Session 1883* (Ottawa: Maclean, Roger, [1883]), 27; Francis Sippley, Baie-Sainte-Anne in Inka Milewski and Janice Harvey, with Sue Calhoun, *Shifting Sands: State of the Coast in Northern and Eastern New Brunswick* (Fredericton, Conservation Council of New Brunswick, 2001), 76. https://conservationcouncilnb.ca/wp-content/uploads/2017/04/Shifting-Sands-2001.pdf.

47 A similar situation has been explored in British Columbia by Douglas C. Harris, *Landing Native Fisheries: Indian Reserves & Fishing Rights in British Columbia, 1849–1925* (Vancouver: UBC Press, 2008). See Maura Hanrahan, "Resisting Colonialism in Nova Scotia: The Kesukwitk [Kepu'kwitk] Mi'kmaq, Centralization, and Residential Schooling," *Native Studies Review* 17, no. 1 (July 2008): 17–44.

48 In Nova Scotia, Puksaqte'kne'katik (Boat Harbour), Fishers Grant, and Merigomish Harbour), Potlotek First Nation (Chapel Island), We'koqma'q (Whycocomagh), and Eskasoni all have saltwater access. The small Sheet

Harbour Reserve (Part of Millbrook First Nation) retains very limited
Atlantic shorefront. Reserve land at Sambro was surrendered to the
Crown in 1919. Indian Island, Ugpi'ganjig (Eel River Bar First Nation);
Esgenoôpetitj (Burnt Church First Nation) in New Brunswick and
Abegweit – Epekwitk and Lennox Island/L'nui Mnikuk in PEI all had
some direct water access through reserve holdings. See Richard H. Bartlett,
Indian Reserves in the Atlantic Provinces of Canada (Saskatoon: University of
Saskatchewan Law Centre, 1986).

49 "R. Williams Blames Fishery Officials," Newspaper and Magazine
Clippings re: fishing industry, clippings, MG 6 A, Vol. 24, 9, RS&Co., NSA.

50 "Pictou Land First Nation Fishing History & Biographies,"
Northumberland Fisheries Museum (NFM), Pictou.

51 Wilson D. Wallis and Ruth Sawtell Wallis, *The Micmac Indians of Eastern
Canada*, (Minneapolis: University of Minnesota Press, 1955), 278.

52 Canada, Department of Citizenship and Immigration, *Report of Indian
Affairs Branch for the Fiscal Year Ended March 31, 1953* (Reprinted from the
Annual Report of the Department of Citizenship and Immigration, pages
40–85 inclusive) (Ottawa: Edmond Cloutier, 1953), 52.

53 By 1983 there were nearly a third fewer lobster licences than there had
been in the early 1970s after 3,000 licences were retired. Gulf Region
Fisheries and Oceans Canada, *A Brief History of the Lobster Fishery in the
Southern Gulf of St. Lawrence Moncton: Gulf Region* (Moncton: DFO, 2012),
31; Gough, *Managing Canada's Fisheries*, 350.

54 For concerns about "Indians" and other local fisheries, see Canada,
"Annual Report Indian Affairs," *Sessional Papers of the Dominion of Canada:
Volume 10, Sixth Session of the Seventh Parliament, Session 1896* (Ottawa:
S.E. Dawson, 1896), oysters 14–46; *Annual Report, Department of Fisheries
1888* (Ottawa: A. Senecal, 1889), salmon and shad, 102, 103; Annual Report
of the Department of M&F, 1895, *Sessional Papers* No. 11A (Ottawa: S.E.
Dawson, 1896), trout, 95; Annual Report of the Department of M&F, 1902,
Sessional Papers No. 22 (Ottawa: S.E. Dawson, 1903), oysters, 261–2. See
Parenteau, "'Care, Control and Supervision.'"

55 Research Material Prince Edward Island Tanya Vance for Boyde Beck,
2009-H0014 Rider Fonds, #5, CMHA; Nicolas Landry, "The Lobster
Industry in the Maritimes 1850–1900," *Les Cahiers* 20, no. 3 (July/September
1989): 104; "Nouvelles de Kent," *Le Courrier des provinces Maritimes*, 28
January 1886, p. 2.

56 Sue Calhoun, *A Word to Say: The Story of the Maritime Fishermen's Union*
(Halifax: Nimbus, 1991), 15; Brun, *La ruée vers le homard*.

57 Brun, *La ruée vers le homard*, 37–40; Régis Brun, "La pêche au homard
en milieu acadien du sud-est du Nouveau-Brunswick, 1850–1900: les
techniques d'exploitation, les entrepreneurs et la main-d'oeuvre." Thèse

de maîtrise en histoire, Université de Moncton, 1988; Ulysse Bourgeois, "Première homarderie du Nouveau-Brunswick," *Les Cahiers de la Société historique acadienne*, 4ᵉ (1962): 4–8.

58 Anselme Chiasson, *Chéticamp: Histoire et traditions acadiennes* (Moncton: Éditions des Aboiteaux,1961), 70–2; Paulette M. Chiasson, "Fiset, Pierre," in *Dictionary of Canadian Biography*, Vol. 13, University of Toronto/ Université Laval, 2003–, accessed 3 September 2019, www.biographi.ca /en/bio/fiset_pierre_13E.html.

59 Rose Mary Babitch, *Le vocabulaire des pêches aux îles Lamèque et Miscou* (Moncton: Les Éditions d'Acadie, 1996), 53.

60 Memo: list of fines in PEI, 18 March 1908, RG23, FO, vol. 284, T 3207, File 2154, Illegal Lobster Fishing 1898–1906, LAC.

61 "Exodus of Fishermen," *Boston Globe*, 26 November 1898, p. 10.

62 Matthew McKenzie, *Breaking the Banks: Representations and Realities in New England Fisheries, 1866–1966* (Amherst: University of Massachusetts Press, 2018).

63 "Port L'Hebert News," *Liverpool Advance*, 10 June 1903, p. 3.

64 Exterior Interpretative panels, 2014, Wallace and Area Museum, Wallace, NS.

65 *An Island Town: A Short History of Clark's Harbour* (np: Archelaus Smith Historical Society, 1984), 8.

66 To Albert J.S. Copp, MP Digby, 31 December 1901, RG23, FO, vol. 296, T 3221, File 2314, Part 6 1901–4, Lobster Regulations, 1900–4, LAC.

Chapter Two

1 Spanier et al., "A Concise Review of Lobster Utilization," 15; Ellen Barlee, *Emigration Papers for the Working Classes: Being a Description of the Climate, Soil, Products, Population, Wages, and General Inducements Offered to Different Classes, of Emigrants in the Various Fields Open to Colonization; New Brunswick* (London: S.W. Partridge; Gordon & Gotch, 18–?), 14, www.canadiana.ca/view/oocihm.48070. Johnston, "The Early Days of the Lobster Fishery," 56–60, demonstrates that lobsters were eaten by settlers.

2 Daniel Samson, *The Spirit of Industry and Improvement: Liberal Government and Rural-Industrial Society, Nova Scotia, 1790–1862* (Kingston and Montreal: McGill-Queen's University Press, 2008), 6. For a discussion of commodities Arjun Appadurai, "Introduction: Commodities and the Politics of Value," in Arjun Appadurai, ed., *The Social Life of Things: Commodities in Cultural Perspective* (Cambridge: Cambridge University Press, 2013), 3–63.

3 Canada, M&F, *Fisheries Statements for the Year 1880* (Ottawa: Maclean, Roger & Co, 1881), 231.

4 Statistics Canada, Table N 83–89, "Value of Exports of Fish and Fish Products, 1868–1975," accessed 27 April 2019, https://www150.statcan .gc.ca/n1/pub/11-516-x/sectionn/N83_89-eng.csv.

5 Canada, *Trade and Commerce, Statistical and Financial Statements 1900* (Ottawa, 1901), 272.

6 In the 1920s, lobster was the only species of concern. Jennifer Hubbard, "The Gospel of Efficiency and the Origins of Maximum Sustainable Yield (MSY): Scientific and Social Influences on Johan Hjort's and A.G. Huntsman's Contributions to Fisheries Science," in *A Century of Maritime Science: The St. Andrews Biological Station*, ed. Jennifer Hubbard, David Wildish, and Robert L Stephenson (Toronto: University of Toronto Press, 2016), 102.

7 Canada, *Annual Report of the Fisheries Branch of the Department of Naval Services for the Year 1915–16* (Ottawa: J. de L. Taché, 1916), 70; *Fifty-Second Annual Report of the Fisheries Branch of the Department of Naval Services, 1918* (Ottawa, 1918), 9; Statistics Canada, Table N83–89: "Value of Exports of Fish and Fish Products, 1868–1975," accessed 6 November 2017, www. statcan.gc.ca/pub/11-516-x/sectionn/N90_100-eng.csv.

8 Korneski, "Development and Degradation"; Kenneth R. Martin and Nathan R. Lippert, *Lobstering and the Maine Coast* (Bath: Maine Maritime Museum, 1985); James Acheson, "The Politics of Managing the Maine Lobster Industry: 1860 to the Present," *Human Ecology* 25, no. 1 (March 1997): 3–27; DFO, *A Brief History of the Lobster Fishery*.

9 Dianne Newell, *The Development of the Pacific Salmon-Canning Industry: A Grown Man's Game* (Montreal and Kingston: McGill-Queen's University Press, 1989).

10 Richard H. Williams, *Historical Account of the Lobster Canning Industry* (Ottawa: Department of M&F, 1930), 3.

11 Richard Rathburn, "The Lobster Fishery," in *The Fisheries and Fishery Industries of the United States, Section V: History and Methods of the Fisheries*, ed. George Brown Goode (Washington, DC: Government Printing Office, 1887), 659.

12 Paul Freedman, "American Restaurants and Cuisine in the Mid-Nineteenth Century," *New England Quarterly* 84, no, 1 (March 2011): 5–59; Andrew Peter Haley, "Turning the Tables: American Restaurant Culture and the Rise of the Middle Class, 1880–1920," (PhD diss., University of Pittsburgh, 2005), 176, 276.

13 Gabriella M. Petrick, "'Purity as Life': H.J. Heinz, Religious Sentiment and the Industrial Diet," *History and Technology* 27, no. 1 (March 2011): 36, 38.

14 "Balancing the Scale," Research Material, 2009–H0014, box 2, Peter Rider Fonds, CMHA.

15 N.D. Jarvis, "Curing and Canning of Fishery Products: A History," *Marine Fisheries Review* 50, no. 4 (Fall 1988): 183.

16 Research Notes, "Balancing the Scale," box 2, page 8, 2009-H0014, Rider Fonds, CMHA.

17 Petrick, "Purity as Life," 37.

18 D.J. McGuirk, Boughton Island, to Temple Macdonald, 14 May 1909, Acc 4225/4/35 Macdonald family [Georgetown] fonds, PARO.

19 Colin McKay, "Co-operation in Lobster Industry," *CF*, August 1922, p. 163.

20 "Affairs in Lobster Industry Discussed," *CF*, April 1923, pp. 91–2. Federal jurisdiction in this area was declared unconstitutional in 1928 as legally caught fish entered provincial jurisdiction once landed (but might re-enter federal control in interprovincial or export trade).

21 Canada, Department of Fisheries, *Annual Report 1936–37* (Ottawa: King's Printer, 1937), 50.

22 A. Gordon DeWolf, *Lobster Fishery of the Maritime Provinces: Economic Effects of Regulations* (Ottawa: Department of Environment, F&M, 1974); M.H. Perley, *Reports on the Sea and River Fisheries of New Brunswick*, (Fredericton: J. Simpson, 1849); Gough, *Managing Canada's Fisheries*, 48.

23 "Prepare for Lent," *Islander* (Charlottetown), 19 March 1858 cited in Boyde Beck, *A Lobster Tale: The Lobster Fishery of Prince Edward Island* (Charlottetown: Prince Edward Island Museum, 2001); Canada, *Seventh Annual Report of the Department of M&F for Year Ending 30th June 1874*, Report of the Commission of Fisheries (Ottawa, 1874), lviii, lx; PEI Lobster research notes, Box 1 #4, 2009-H0014, Rider Fonds, CMHA; Douglas Baldwin, *Prince Edward Island: An Illustrated History* (Halifax: Nimbus, 2009), 203.

24 George A. McInnis, "A Brief History of the Lobster Canning Industry," *Vanguard*, 30 April 1969, in Scrapbook Fonds Père Clarence d'Entremont, Série Q, Scrapbook, Musée des Acadiens des Pubnicos et Centre de Recherche, West Pubnico, NS.

25 Williams, *Historical Account of the Lobster Canning Industry*, 6. See also Korneski, "Development and Degradation," 28.

26 "The Lobster Fishery," *Shelburne Coast Guard*, 15 December n.d., #2 Scrapbook of newspaper clippings on the Lobster Industry, MG6 A, vol. 24, RS&Co., NSA; Rathburn, "Lobster Fishery," 688.

27 James M. Acheson and Roy Gardner, "The Evolution of Conservation Rules and Norms in the Maine Lobster Industry," *Ocean and Coastal Management* 53, no. 9 (2010): 535; Acheson, *Lobster Gangs*, 4; Judd, "Saving the Fisherman"; Bolster, *The Mortal Sea*, 212.

28 Rathburn, "Lobster Fishery," 687; United States Senate Committee on Finance, *Rates of Duty on Imports to the United States from 1789 to 1890, Inclusive* (Washington: Government Printing Office, 1890), 836; United States, Tariff Commission, Underwood Tariff (New York: C.S. Hammond, 1913), 5.

29 "Sambro Fish Establishment," *Halifax Citizen*, 26 November 1863.
30 Edward MacDonald, *If You're Stronghearted: Prince Edward Island in the Twentieth Century* (Charlottetown: Prince Edward Island Museum and Heritage Foundation, 2000), 26–7. See also Nancy Gorveatt, "'Polluted With Factories': Lobster Canning on Prince Edward Island," *Island Magazine* 57 (Spring/Summer 2005): 10–21.
31 H.E. Baker, FCS Draft, 10th Article; "Early American Canners At Home and Abroad," 5; Canadian Lobster Fishery Statistics and other data from 1869 to 1896, RS&Co., MG6 A, vol. 24, 10, NSA.
32 Alex Gunn to R.H. Williams, 12 March 1930, Canadian Lobster Fishery Statistics and other data from 1869 to 1896, RS&Co., MG6 A, vol. 24, 10, NSA.
33 Canada, M&F Committee, *Evidence Taken Before the M&F Committee Respecting the Lobster Industry During the Session of 1909* (Ottawa: C.H. Parmalee, 1909), *xxix–xxxii*; Canada, *Canadian Lobster Commission*, 20. This spectacular rate of growth in part characterized many new resource industries in Canada. See: "National Economic Patterns," in *Historical Atlas of Canada: Addressing the Twentieth Century*, ed. Cole Harris and Geoffrey J. Matthews (Toronto: University of Toronto Press, 1987), 13–15.
34 *Evidence Taken*, 15.
35 See Morton, "Alien Concerns." Almost all canned fish in the Maritimes was lobster. Although much of the herring caught for Maine's sardine processing came from Canadian waters, before 1900 almost all sardine packing took place there. Thereafter, sardine packing expanded north of the border to Black Harbour, Deer, and Grand Manan Islands. Ross Coen, "Selling Salmon to the World: The Export Market for Pacific Northwest Canned Salmon," *The Pacific Northwest Quarterly* 105, no. 1 (2013): 23–31; Brian Payne, "Becoming a Dependent Class: Quoddy Herring Fishermen in the 1920s," *Labour / Le Travail* 81 (Spring 2018): 87–117; "Report of the Deputy Minister," Fisheries Report, *Annual Report of the Department of M&F 1901* (Ottawa: S.E. Dawson, 1902), xxix.
36 Beck, *A Lobster Tale*.
37 Research Notes, "Balancing the Scale," 17, box 2, theme 6, Lobster, Rider Fonds, CMHA.
38 For example, see manuscript of A.J. Macdonald memoirs, 1909, Acc 4225/5/1/42, Macdonald family [Georgetown] fonds, PARO.
39 H.N. Morse, "Section N: Fisheries," Historical Statistics of Canada, accessed 13 November 2017, www.statcan.gc.ca/pub/11-516-x/sectionn/4057755-eng.htm.
40 William Templeman, *The Newfoundland Lobster Fishery; an Account of Statistics, Methods and Important Laws*, Research Bulletin No. 11 (St. John's Department of Natural Resources, 1941), 9.

41 FO Canada, "Lobster," accessed 13 November 2017, www.dfo-mpo.gc.ca /fm-gp/sustainable-durable/fisheries-peches/lobster-homard-eng.htm.

42 *Evidence Taken*, 54, 231.

43 Jacquard, *Lobstering*, 43.

44 8 May 1906 agreement between Alfred Morrison and A. & R. Loggie, Merchants of Loggieville, NB, A&R Loggie Fishing Licenses and Contracts, 1866–1908, MG28, III, 13, LAC.

45 Jacquard, *Lobstering*, 59.

46 See report of W.H. Rogers, Fishery Officer, *Report of the Commissioner of Fisheries for the Year Ending 31 December 1877* (Ottawa: Maclean, Roger & Co, 1878), 158.

47 James P. Baxter, Sr. to W.F. Tidmarsh, 29 December 1918; Research Notes – Cannery, MS series 4, box 1, folder 1, Maine Maritime Museum. Bath, ME; Arsenault, "Packing Lobsters at the Beach"; Korneski, "Development and Degradation," 29–30. Paul B. Frederic, *Canning Gold: Northern New England Sweet Corn Industry: A Historical Geography* (Lanham, MD: University Press of America, 2002), 28–9.

48 Arthur I. Judge, ed., *A History of the Canning Industry by Its Most Prominent Men* ([n.p.]: National Canners Association, 1914), 41. Albert Nelson Marquis, *Who's Who in New England; a Biographical Dictionary of Leading Living Men and Women of the States of Maine, New Hampshire, Vermont, Massachusetts, Rhode Island and Connecticut; New England Manufacturers and Manufacturies: Three Hundred and Fifty of the Leading Manufacturers of New England*, vol. 2 ([n.p.]: Van Slyck, 1879), 764. River John Community Access Program, "American Lobster," 2002, www.parl.ns.ca/lobster /history.htm#The%20Very%20Beginning (accessed 9 November 2017); Bourgeois, "Premières homarderies du N.-B.," 5.

49 *Evidence Taken*, 196. H.C. Baxter Brothers sold its last lobster factory in Canada to Burnham and Morrill in 1916. See Baxter to Tidmarsh, 29 December 1918.

50 W.F. Tidmarsh to Louis-Phillippe Brodeur, 7 February 1908; Tidmarsh to Brodeur, 8 February 1908; RG23, FO, vol. 297, T-3222, file 2314, parts 8, Lobster Regulations, 1906–9; William S. Fielding to Raymond Préfontaine, 5 April 1905, RG23, FO, vol. 296, T-3222, file 2314, part 7: Lobster Regulations, 1904–6, LAC.

51 Fielding to Préfontaine, 5 April 1905.

52 Tidmarsh to Brodeur, 7 February 1908; Tidmarsh to Brodeur, 8 February 1908; Tidmarsh to Myrick cited in MacDonald, *If You're Stronghearted*, 172. The same year, a Department of Fisheries inspector in south-west Nova Scotia reported on a combine involving seven American, British, and Canadian packers to control product and corner the can and twine market. "Report – Combine to Control Lobster Industry 1913," RG23, FO,

vol. 408, T-3396, file 4665, part 1, LAC; Great Britain, *Royal Commission on the Natural Resources, Trade, and Legislation of Certain Portions of His Majesty's Dominions: Minutes of Evidence Taken in the Maritime Provinces of Canada in 1914* (London: Eyre and Spottiswoode, 1915), 137.

53 Ward Fisher, Inspector to Found, 19 November 1913, RG23, FO, vol. 408, T 3396, File 4665, Part 1, Report – Combine to Control Lobster Industry 1913, LAC. Those in attendance were Burnham and Morrill, Roberts, Simpson and Company, O'Leary & Lee, the Neville Company, the Smiths, the Bates and the Cann Packing firms. For a parallel discussion of the trust issue in the sardine industry see Payne, "Becoming a Dependent Class."

54 To J.W. Smith, 16 December 1912, Correspondence, minutes, clippings. 1913–34, MG6 A, vol. 24, 2 RS&Co., NSA.

55 See, for example, 30 April 1909, G.J. Desbarats, Acting DM M&F to Donald Scott, Fishermen's Union, Main-à-Dieu, RG23, FO, vol. 297, T 3981, File 2314, Part 9, Lobster Regulations 1909–10, LAC.

56 *Evidence Taken*, 211; M. Neville to Brodeur, Minister M&F, 4 February 1907, RG23, FO, vol. 297, T 3222, file 2314, part 8, Lobster Regulations 1906–9, LAC.

57 Alfred Morrison fishing contract; See also: H.W. Longworth to Senator Donald Ferguson, 15 October 1895, RG23, FO, vol. 294, T 3219, file 2314, part 1, Lobster Regulations 1895–6, LAC.

58 Frederic, *Canning Gold*, 82.

59 "Who's Who in the Fishing World," *CF*, July 1915, p. 208; Princes Group owned by the Mitsubishi Corp. since 1989 offers brands such as Crosse & Blackwell.

60 "History of the Lobster Industry in the Northumberland Strait," *Pictou Advocate*, 12 July 1934.

61 *The Mercantile Agency Reference Book and Key, 1887* (Montreal: Dun Wiman, 1887); Kennedy Wells, *The Fishery of Prince Edward Island* (Charlottetown, PEI: Ragweed Press, 1986); Arsenault, "Packing Lobsters at the Beach," 1; MacDonald, *If You're Stronghearted*, 172. J. Winslow Jones went bankrupt in 1881, was reorganized as the Winslow Packing Company, and was later purchased by the Portland Packing Company. Frederic, *Canning Gold*, 44.

62 Stéphanie Arseneau Bussières and Hélène Chevrier, *Socio-Economic Profile of the Magdalen Islands' English-Speaking Community* (Havre-aux-Maisons, QC: Centre de recherché sur les milieu insulaires et maritime, 2008), 6. www.uqar.ca/uqar/recherche/unites_de_recherche/cermim/english _community.pdf, accessed 3 September 2019.

63 Rathburn, "Lobster Fishery," 690.

64 Canada, *Census* (1890–1), vol. 3 (Ottawa: S.E. Dawson, 1894), 149–50. In July 1886, the Burnham and Morrill factory at Tor Bay on Nova Scotia's Eastern Shore burned down, with losses (including 800 cases of lobster)

estimated at $15,000. See "Messrs. Burnham & Morrill's Lobster Factory," *Liverpool Advance*, 14 July 1886. The New Brunswick numbers would also include the sardine industry.

65 Canada, *Report of the Lobster Industry* (Ottawa: S.E. Dawson, 1892), 31.
66 See, for example, Canada, Lobster and Oyster Commission, *Report*, 11.
67 James A. Fraser, *A History of the W.S. Loggie Co Ltd, 1873–1973* (Fredericton: Provincial Archives of New Brunswick, 1973).
68 Ian Sclanders, "N.B.'s Noble Lobster," *Maclean's*, 1 January 1952, 35.
69 S. Taylor, Bonne Bay to Walter Skanes, 4 April 1930; Skanes to R.A Squires, 4 April 1930, GN8, vol. 143, Office of the Prime Minister, Lobster Regulations, PANL.
70 William I. Payne to Skeans, 29 March 1930, GN8, vol. 143, Office of the Prime Minister, Lobster Regulations, PANL.
71 Arsenault, "Packing Lobsters at the Beach," 13.
72 Babitch, *Le vocabulaire des pêches*; Ranald Thurgood, "Storytelling on the Gabarus-Framboise Coast of Cape Breton: Oral Narrative Repertoire Analysis in a Folk Community," (PhD diss., Memorial University of Newfoundland, 1999), 173; Scrapbook Edna Thompson Brown, 2000 interviews, NFM.
73 Canada, *Report of the Royal Commission on the Relations of Labor and Capital in Canada* (Ottawa: A. Senecal, 1889), 69.
74 Arsenault, "Packing Lobsters at the Beach," 13. Leo Cheverie, "Johnston's Lobster Factory, 1935–1945," Undergraduate paper, 1982, University of PEI Archives, Charlottetown, p. 11.
75 "Le Homard," *Moniteur Acadien*, 10 August 1886, p. 2.
76 Régis Brun, "Le développement inégal de la pêche acadienne du homard au sud-est du N.-B., 1880–1930: la naissance des Coopératives de pêcheurs," 1985, tableau 6: "Tableau comparatif des salaires de métiers des ouvriers aux Maritimes c. 1886." No. 101.95, Fonds Régis Brun, Homard- Thèse et Articles de pérodique, Centre d'études acadiennes Anselme-Chiasson, Université de Moncton, Moncton.
77 "Barrington Notes," *Yarmouth Herald*, 31 March 1881, n.p.
78 Wakeham, Report on Lobster Investigation c 1909, RG23, FO, vol. 323, T-4008, File 2750, Part 1 1898–9, Lobster Commission 1898–1910, LAC.
79 "Sambro Fish Establishment," *Halifax Citizen*, 26 November 1863; Ruth Fulton Grant, *The Canadian Atlantic Fishery* (Toronto: Ryerson Press, 1934), 28–9; Williams, *Historical Account of the Lobster Canning Industry*.
80 Notes and interviews recorded by Richard Rathburn, 27 May 1895, p. 9, RG22 US Fish and Wildlife Service Office of the Commissioner of Fish and Fisheries, P44 Records of the Joint Fisheries Commission Relative to the Preservation of the Fisheries in Waters Contiguous to Canada and the US, 1893–5 Container #20, Joint Fisheries Commission Inquiries,

Mackerel NS Coast May and June 1895, National Archives and Records Administration, College Park, MD.

81 "To Fishermen," *Yarmouth Herald*, 26 June 1899, n.p.

82 A.R. Gordon, "Lobster Fishing Industry," Canada, *Sessional Paper* 1891, vol. 8 appendix 8, 127–39; Brun, "Le développement."

83 Memo, Found, 9 November 1911, RG23, FO, vol. 298, T 3981, File 2314, Part 10A, Lobster Regulations – Minister's File, 1912, LAC.

84 M.A. Nickerson, Clark's Harbour to Charles Petite, Yarmouth, 21 November 1928, MG6 A, vol. 24, 2 Correspondence, minutes, clippings. 1913–34; Moses H. Nickerson, Melrose, MA letter, 27 January 1930, MG6 A, vol. 24, 10, Canadian Lobster Fishery Statistics and other data from 1869–96. RS&Co., NSA.

85 "Lobster Combines' Queer Document," *Advance*, 27 December 1905; Marion Robertson, "Robertson, Thomas," in *Dictionary of Canadian Biography*, vol. 13, University of Toronto/Université Laval, 2003–, accessed 13 July 2020, www.biographi.ca/en/bio/robertson_thomas_13E.html.

86 "With Wilfred Poirier. Lobster Buyer," *Cape Breton Magazine* 33 (June 1983): 53–9.

87 Baldwin, *Prince Edward Island*, 168.

88 "Canned Lobster Industry Feel Jap Competition," clipping n.d., MG6 A, vol. 24, 10 Canadian Lobster Fishery Statistics and other data from 1869–96. RS&Co., NSA.

89 "Prince Edward Island Fisheries," *CF*, April 1921, 90; "Poor Outlook for Canned Lobster," *CF*, December 1923, 323.

90 *Annual Report of the Newfoundland Department of Fisheries for the Year 1911*, xxiv; Frederic, *Canning Gold*, 82; *Pictou, Nova Scotia, Canada: Nova Scotia's Northern Ocean Port* (Pictou, NS: Town Council of Pictou, 1916), 49, 51; Liam Mather, "W.F. Tidmarsh and Lobster Canning in Prince Edward Island: A Social Biography," undergraduate paper, McGill University, 2016 (in possession of author).

91 Hubbard, *Science on the Scales*, 112–17.

92 "Eastern Canadian Fishery Conference," *CF*, October 1921, 227. A detailed description of proper lobster canning appeared in Ernest Hess, "Lobster Canning," *Canadian Canner and Food Manufacturer* April 1935, pp. 9–11.

93 J.W. Smith, "Notes on the Lobster Trade" *Canadian Food Packer*, June 1938, MG6 A, vol. 24, no. 9, Newspaper and Magazine Clippings re: fishing industry. RS&Co., NSA.

94 "Brine Freezing of Lobster," *CF*, May 1929, p. 26; "Austin E. Nickerson Ltd Pack Brine Frozen Lobster," *CF*, July 1930, p. 20. Émile Paturel is often given credit for this for this innovation in 1936. See Cathy Billings, *The Maine Lobster Industry: A History of Culture, Conservation & Commerce* (Charleston, SC: History Press, 2014), 51.

95 *The Lobster Industry* (Pictou: Hector Centre Trust, 1980), 9.

96 Maine, *5th Report of the Commission on Sea and Shore Fisheries,*1928 (NP: Rockland Maine), 13.

97 Maureen Larkin, "Our Way of Living: Survival Strategies in Lobster Fishing Households in PEI," (MA thesis, Memorial University 1990), 39.

98 Judd, "Saving the Fisherman:" 619.

99 "Trial Shipment of Lobsters Leaves for Boston by Trucks," clipping, 20 August 1931, MG6 A, vol. 26, No 1 Scrapbook of newspaper clippings etc. on the Lobster industry. RS&Co., NSA.

100 "The Carferry Meets the Car: The S.S. PEI and the Auto," Sailstrait, accessed 20 July 2022, https://sailstrait.wordpress.com/2015/09/29 /the-carferry-meets-the-car-the-s-s-p-e-i-and-the-auto/; "The Lobster Industry," *CF*, August 1933, pp. 16–17.

101 Harold Russell, b. 1929 father General Manager, p. 43, interviews 2000, NFM.

102 Nonnie Murray, "What William Broidy Did for Nova Scotia," *Chronicle*, 1 April 1972, clipping, NFM.

103 "A Look into the Lobster Industry's Past," Fisheries Museum Research project 1985, NFM, p. 8; "National Sea buys Two Lobster Companies," *CF*, May 1965, p. 8.

104 "Factories along the Northumberland Strait," NFM; Clipping, "A Notable Man of Many Interests," *The Busy East*, n.d., 30–2, MG4, vol. 7, file 14, no. 25, North Shore Archive Society, Tatamagouche Heritage Centre. "The Man Behind Mephisto: Fred Magee, an Extraordinary Entrepreneur," Tantramar Heritage Trust, June 2006, accessed 13 July 2020, https:// tantramarheritage.ca/2002/06/the-man-behind-mephisto-fred-magee -an-extraordinary-entrepreneur/.

105 Russell interview, p. 42-3; Olive Prest Pastor, ed., *Stories of Caribou* (Caribou District Two Heritage Society, 1992), 169.

106 "Emile Paturel décédé hier," *L'Évangéline*, 9 March 1950, p. 7; Fonds André-Paturel, Centre d'études acadiennes Anselme-Chiasson, accessed 13 July 2020, www.umoncton.ca/umcm-ceaac/files/umcm-ceaac/wf /wf/pdf/106-AndrePaturel.pdf.

107 "Lobster Situation 'Reasonably Satisfactory,'" *Herald*, 7 July 1934, scrapbook H.G. Connor, Fisheries Museum of the Atlantic, Lunenburg.

108 Lee Mawsley and Kay MacIntosh, "A History of Merigomish," undergraduate paper, St. Francis Xavier, 1972, NFM; poster Museum Exhibit, NFM.

109 Lobster Packing Licenses, MG6 A, vol. 24, 4, Meeting of the Maritime Canned Fish Section 1940 Amherst, Minutes. RS&Co., NSA.

110 Patrick Sauer, "When Disaster Strikes," Inc.com, August 2008, www.inc .com/articles/2008/08/lobster.html; Ian Sclanders, "Lobster Ranch," *Maclean's*, 15 December 1946, 16.

111 "Nova Scotian Woman Has Revolutionized Lobster Industry," 76 c.
    1935, MG6 A, vol. 26, no.1, Scrapbook of newspaper clippings etc. on the
    lobster industry, RS&Co., NSA; Crie to Mrs. Fannie Powell, 2 March 1931,
    Sea and Shore Commissioner's Correspondence, 1/1/1930–1/31.1931
    Box 43, Maine State Archives, Augusta; "Nova Scotia Woman Runs Big
    Lobster Business Daughter of Moses Nickerson," CF, March 1936, p. 10;
    Jacquard, Lobstering, 236–7.
112 "Good Season on Eastern Shore," Halifax Herald, 18 August 1931,
    scrapbook H.G. Connor, Fisheries Museum of the Atlantic, Lunenburg.
113 "Big Lobster Cargo," Oxford Journal, 18 May 1933, p. 1. See also A.F.
    Chaisson, "Factors in the Shipment of Live Lobsters from Eastern Nova
    Scotia," Bulletin XXXIII (Ottawa: Biological Board of Canada, 1932).
114 "Conley's New Pound," CF, May 1932, p. 13; "Conley Extends Live
    Lobster Pounds at Deer Island," CF, May 1935, p. 21; "National Sea Buys
    Conley's Lobsters," CF, August 1965, p. 37.
115 "Lobster Price Is Way Down," Oxford Journal, 21 April 1932, p. 1.
116 Sclanders, "Lobster Ranch," 24.
117 Clipping, Halifax Chronicle, 10 December 1931, MG6 A, vol. 26, no. 1
    Scrapbook of newspaper clippings etc. on the lobster industry, RS&Co.,
    NSA; Judd, "Saving the Fisherman:" 618; Gordon McKean, "Lobstermen
    in Million Dollar Race," Halifax Herald, 3 May 1937, p. 8; There were
    13,188 licences in 1929, and this rose to a peak of 18,011 in 1936, as few
    other employment options were available. DeWolf, Lobster Fishery of the
    Maritime Provinces, 25.
118 Cheverie, "Johnston's Lobster Factory, 1935–1945," 3; "Fisherfolk Existing
    in Direst Want," Halifax Mail, 4 July 1934, scrapbook H.G. Connor,
    Fisheries Museum of the Atlantic, Lunenburg, NS.
119 Dennis Point café placemats, Lower West Pubnico, NS, 2019.
120 Jacquard, Lobstering, 335.
121 "Lobster Fisherman, Son Drown," Halifax Herald, 27 May 1937, p. 1.
122 "Lobster Plug Story in West Pubnico," 20 April 1981, Clippings, Library
    Fisheries Museum of the Atlantic, Lunenburg, NS.
123 "Clark's Harbor Community Notes Column," Cape Sable Advertiser, 7
    October 1886.
124 Canada, Lobster Industry, 1.
125 Minutes, MG6 A, vol. 24, 4 Meeting of the Maritime Canned Fish Section
    1940 Amherst, Minutes, RS&Co., NSA.
126 Payne, Eating the Ocean, 153.
127 Catherine Caldwell Bayley, "It's Patriotic and Pleasant to Eat Canadian
    Lobster," Canadian Home Journal 37, no. 3 (July 1940): 28–9 and Canadian
    Home Journal 36, no. 8 (December 1939): 1; "Canada Easts More Canned
    Lobster," CF, June 1941, p. 8; Canada, Department of Fisheries 1941–42
    (Ottawa: Edmond Cloutier, 1942), 11.

128 DeWolf, *Lobster Fishery of the Maritime Provinces*, 26.

129 Canada, *Canada Year Book 1960* (Ottawa: Dominion Bureau of Statistics, 1960), 1030–1.

130 "National Sea Buys Conley's Lobsters," *CF*, August 1965, p. 37; "Flying Lobsters by Air," *CF*, January 1946, p. 34; "Ship Lobster by Air," *CF*, May 1946, p. 30; "Shipping Cooked Lobster," *CF*, July 1946, p. 39; Andrew Coe, *Chop Suey: A Cultural History of Chinese Food in the United States* (New York: Oxford University Press, 2009); "Getting Closer to Your Food," Restaurant-ing through History, 29 April 20201, https://restaurant-ingthroughhistory.com/tag/lobster-tanks/; "The Origins of Chinese Lobster Sauce," accessed 19 March 2023.

131 "Pictou Hails King Lobster," *CF*, August 1937, p. 9; Jacquard, *Lobstering*, 154.

132 Research notes, 2009–H0014, Rider Fonds "Balancing the Scales," Box 2, p. 3, CHMA; New Glasgow Lobster Suppers, https://peilobstersuppers.com/history/ accessed 26 April 2021; "If You Love Lobster, Welcome to Paradise," *Maclean's*, 27 April 1981, https://archive.macleans.ca/article/1981/4/27/if-you-love-lobster-welcome-to-paradise.

133 Russell interview, p. 4, NFM.

134 Maritime Lobsters Ltd., to Goss, DM, 12 November 1962, GN 34, Department of Fisheries, Box 10, File /387, Maritime Lobsters Ltd (A.W. Swimm), PANL.

135 "National Sea Buys Two Lobster Companies," *CF*, May 1965, p. 8.

136 Sclanders, "Lobster Ranch," 24; "Shortage of Boats for Lobster Season Opening," *CF*, May 1946, p. 25.

137 L. Richard Lund, "'Fishing for Stamps': The Origins and Development of Unemployment Insurance for Canada's Commercial Fisheries, 1941–71," *Journal of the Canadian Historical Association* 6, no. 1 (1995): 185.

138 Canada, *Year Book* 1955, 581.

139 "Stern Warning Issued to Lobster Poacher," *CF*, November 1958, p. 22; Urbain LeBlanc, "The Lobster-Trap Population Explosion," *CF*, June 1964, p. 67.

140 Research notes, 2009–H0014, Box 2, p. 17, Rider Fonds "Balancing the Scales," CMHA.

141 A.A. MacDonald and W.B. Clare, *Guysborough Shore Resource Survey*, 1966, viii, RG30–3/1/ 6796, Extension Department, Fisheries, St. Francis Xavier University Archives (SFXUA).

142 Alexander Laidlaw to A.L Barry, Director of Eastern Fisheries, 18 May 1950; Laidlaw to Stewart Bates, 10 May 1950, RG 30–3/1/ 4415–25; 4426–40 Extension Department, Fisheries, SFXUA.

143 DeWolf, *Lobster Fishery of the Maritime Provinces*, 20; Munro McKenzie, River John to Minister M&F, 28 May 1912, RG23, FO, vol. 298, T 3981, File 2314, Part 11, Lobster Regulations 1912–14, LAC.

144 No legal limit on the number of traps that could be fished was introduced until 1966 when fishermen in District 8 were limited to 250 traps. The following spring fishermen on the north side of PEI could fish no more than 500 traps. Canada, *Lobster Fishery Task Force, Final Report* (Ottawa: DFO, 1975), 63; Andrew Swimm to Maloney, assistant DM, 28 May 1963, GN34, Department of Fisheries, Box 10, File /387, Maritime Lobsters Ltd. (A.W. Swimm), PANL.

145 Caitlin Nicole Johnson, "Voices of the North Shore: The Lobster Fishery of Pictou County," (MA thesis, Saint Mary's University, 2013), 9 and her chapter 6.

Chapter Three

1 Canada, *Canadian Lobster Commission 1898*, 1, 28, 23; "Eastern Canadian Fishery Conference," *CF*, October 1921, p. 227.

2 Canada, M&F Committee, *Evidence Taken Before the M&F Committee Respecting the Lobster Industry During the Session of 1909* (Ottawa: C.H. Parmalee, 1909), R.N. Venning, 122.

3 Ronald D. Tallman, "Peter Mitchell and the Genesis of a National Fisheries Policy," *Acadiensis* 9, no. 2 (Spring 1975): 66–78.

4 Cadigan, "Moral Economy of the Commons;" Karl Jacoby, *Crimes Against Nature: Squatters, Poachers, Thieves, and the Hidden History of American Conservation* (Berkeley: University of California Press, 2014 ed.).

5 Payne, "Local Economic Stewards," 31.

6 For a context on centralized decision-making and the weakness of the state, see Jean-Marie Fecteau, *The Pauper's Freedom: Crime and Poverty in 19th Century Quebec* (Montreal and Kingston: McGill-Queen's University Press, 2017), 100.

7 K.W. Hewitt, "The Newfoundland Fishery and State Intervention in the Nineteenth Century: The Fisheries Commission, 1888–1893," *Newfoundland Studies* 9, no. 1 (Spring 1993): 58–80; Bill Parenteau and Richard Judd, "More Buck for the Bang: Sporting and the Ideology of Wildlife Management in New England and the Maritime Provinces, 1870–1900," in *New England and the Maritime Provinces: Comparisons and Connections*, ed. Stephen Hornsby and John Reid (Montreal & Kingston: McGill-Queen's University Press, 2005). For economic interests, see Judd, "Saving the Fisherman;" Arthur McEvoy, *The Fisherman's Problem: Ecology and Law in the California Fisheries, 1850–1980* (Cambridge: Cambridge University Press, 1986); Richard A. Cooley, *Politics and Conservation: The Decline of Alaska Salmon* (New York: Harper & Row, 1963).

8 William E. Schrank, "Extended Fisheries Jurisdiction: Origins of the Current Crisis in Atlantic Canada's Fisheries," *Marine Policy* 19, no. 4 (July

1995): 285–99. See also Brian Payne, *Fishing a Borderless Sea: Environmental Territorialism in the North Atlantic, 1818–1910* (Ann Arbor: University of Michigan Press, 2010); Gough, *Managing Canada's Fisheries*.

9 RG22, US Fish and Wildlife Service, Dept. of Commerce, Bureau of Fisheries, P4, Records concerning relations with Canada, 1905–36, Container 14, Atlantic Coast Fisheries 1907–18, National Archives, Washington. The most important case became an international incident in 1911 when two Boston-owned vessels, the *J. R. Atwood* and *Pride of the Port*, fished off Seal Island, Shelburne County, during a close season. To Minister M&F, 27 March 1911, RG23, FO, vol. 298, T 3981, File 2314, Part 10A, Lobster Regulations – Minister's File, 1912; William Smith, DM M&F, to E.L. Newcombe, DM Justice, 9 September 1895; handwritten memo, 23 September 1895; "M&F – Lobster Caught Outside the Three-Mile Limit: 1895–1895," RG13, Department of Justice, Series A-2, vol. 99, File 1895–838, File Title: /10; "If Lobster Fishermen May Set Their Gear Outside the 3-Mile Limit Before Open Season Begins," from Department of M&F, 1910–13, RG13, Justice, Series A-2, vol. 2365, File 1913–691, LAC.

10 Harris, *Fish, Law and Colonialism*; Harris, *Landing Native Fisheries*, 8; Geoffrey Moyse, "Untangling Some of the Tangled 'Lines' Federal/ Provincial Constitutional Jurisdictions over Fisheries," *The Advocate* 50, no. 6 (1992): 939; Bonnie J. McCay, *Oyster Wars and the Public Trust: Property, Law and Ecology in New Jersey History* (Tucson: University of Arizona Press, 1998); *Fisheries Reference Attorney General Canada v. Attorney General Ontario (Fisheries)* [1898] AC 700, Judicial Committee Privy Council, 9.

11 A distinct Department of Fisheries was resurrected in 1930 and continued until 1969. Loran E. Baker memo, n.d. RG30–3/1/ Extension Department, 232–61, Fisheries; SFXUA. For a discussion on commons, see Daniel W. Bromley, "The Commons, Property, and Common-Property Regimes," in *Making the Commons Work: Theory, Practice, and Policy*, ed. Daniel W. Bromley et al. (San Francisco: ICS Press, 1991), 3–15; Ostrom, *Governing the Commons*; Thompson, *Customs in Common*; Thompson, *Whigs and Hunters* (New York: Pantheon, 1975); James C. Scott, *Seeing Like a State: How Certain Schemes to Improve the Human Condition Have Failed* (New Haven: Yale University Press, 1998).

12 Gough, *Managing Canada's Fisheries*, 158.

13 Parsons, *Management of Marine Fisheries*, 22; Hubbard, *Science on the Scales*, 93.

14 Canada, "Special Report Regulation Affecting Lobster Fishery," 1873, 153.

15 This position was held so that that the French had no claim to exclusive rights along the French shore. Sharon A. Firestone, "To Plant with Women: Women's Lives in Savage Cove, Newfoundland," (PhD diss., Arizona State University, 2003), 67.

16 John Louis Babin, Caplin River, Bonaventure, to Department of M&F, 23 March 1911, RG23, FO, vol. 298, T 3981, File: 2314, Part 10A, Lobster Regulations – Minister's File, 1912, LAC.

17 McEvoy, *Fisherman's Problem*; Martin & Lippert, *Lobstering and the Maine Coast*, 13, 43, 45; Philip Conkling and Anne Hayden, *Lobsters Great and Small: How Scientists and Fishermen Are Changing Our Understanding of a Maine Icon* (Rockland, ME: Island Institute, 2002), 6–7. Conservation laws from the 1850s had restricted access to the lobster fishery to local residents. Kevin H. Kelly, "A Summary of Maine Lobster Laws and Regulations 1820–1990," Lobster informational leaflet #19, Lobster Reference Collection, Box 3 Maine Maritime Museum Library, Bath, ME.

18 Judd, "Saving the Fisherman."

19 Conkling and Hayden, *Lobsters Great and Small*, 7.

20 Joseph William Collins, *Report upon a Convention Held at Boston, 1903, to Secure Better Protection of the Lobster* (Boston: Wright and Porter, 1904).

21 "Prince Edward Island Fisheries," *CF*, March 1920, p. 69; Canada, Order in Council 0842, 7 July 1873; Order in Council 0437, 23 April 1874; Order in Council 0495, 26 May 1877; Order in Council 0348, 13 March 1879; Order in Council 2558, 7 December 1899. See also R.H. Williams, "Lobster Report Presents Many Problems," *CF* October 1928, p. 21, which includes a season chart noting ice, moulting, and spawning information also with demand, quality, supply, and quality of product.

22 DM M&F to Robert Bell, Waterford, PEI, 14 June 1899, RG23, FO, vol. 295, T 3220, File 2314, Part 4, Lobster Regulations 1898–1900, LAC.

23 A. Brooker Klugh, "Saving the Lobster Industry," *CF*, May 1918, pp. 731–3; G.J. Desbarats, Correspondence, *CF* June 1918, pp. 773–4.

24 These same canners were in an almost constant campaign to shut down these same small canners, as they supposedly undermined the quality of Canadian canned lobster. M.H. Nickerson, "Alleged Lobster Decline," *CF*, June 1918, pp. 775–6; "Prince Edward Island Notes," June 1918, p. 797.

25 "Antidote for Lobster Smut Found," *CF*, February 1923, p. 33; "Affairs in Lobster Industry Discussed," *CF*, April 1923, pp. 91–2.

26 "Editorial, "The Consolidation of Lobster Canneries," *CF*, September 1923, p. 225; "The Lobster Industry," *CF*, January 1935, p. 18.

27 Clipping "Newfoundland Fishery: Marked Success in Propagating Cod and Lobster," *New York Tribune*, 6 December 1891; Memo Prince, 23 April 1903, RG23, FO, vol. 290, T 3214, File 2208, Part 1, Patent – Lobster Incubators, 1895–1905, LAC.

28 Edward E. Prince, "Fifty Years of Fishery Administration in Canada," *Proceedings of the American Fisheries Society 1920 Meeting in Ottawa* 50, no. 1, https://doi.org/10.1577/1548-8659(1920)50[163:FYOFAI]2.0.CO;2.

29 1894 Bill with handwritten revisions amendments to Section 10, RG23, FO, vol. 253, T 3173, File 1645; Lobster Regulations, 1894–1905, LAC.

30 F. Gourdeau, DM M&F to A.B. Crosby & Co, Halifax, 1 April 1899, RG23, FO, vol. 288, T 3213, File 2186, Parts 1, Lobster Leases – Labrador, 1895–1905, LAC.

31 Brief on the Lobster Industry submitted to Hon. J. Angus MacLean Minister of Fisheries, November 1962 by Fisheries Extension Committee of STFX, p. 4, RG30–3/1/ Extension Department, 42, Fisheries, SFXUA.

32 I am grateful to Ed MacDonald for this point.

33 From "Charlottetown Merchant," 28 August 1894, RG23, FO, vol. 253, T 3173, File 1645; Lobster Regulations, 1894–1905, LAC.

34 A.C. MacDonald to Charles Tupper, 8 September 1894, RG23, FO, vol. 253, T 3173, File 1645; Lobster Regulations, 1894–1905; To Wm Smith, DM M&F from H.E. Baker, 17 September 1894 and Department Curricular, 17 September 1894, RG23, FO, vol. 253, T 3173, File 1645; Lobster Regulations, 1894–1905, LAC.

35 Poster dated 1 April 1923 is located and displayed at the Wallace and Area Museum, Wallace Nova Scotia.

36 Memo re: having lobster in possession during the close season, 1894, RG23, FO, vol. 243, T 3163, File 1551, Shipment of Illegally Caught Lobster to United States, 1894–1913, LAC.

37 Memo: Impounding Lobsters in Close Season by Prince, 25 May 1896; Memo: Impounding Lobsters in Close Season, 20 February 1896, RG23, FO, vol. 243, T 3163, File: 1551, Shipment of Illegally Caught Lobster to United States, 1894–1913, LAC.

38 Memo from Prince, 25 May 1895, RG23, FO, vol. 270, T 3193, File 1901, Lobster Ponds and Hatcheries, 1894–1909, LAC.

39 Prince to L.S. Ford, Inspector, 7 August 1897; Ford to Prince, 1 October 1897; Prince to Ford 10 November 1897; RG23, FO, vol. 270, T 3193, File 1901, Lobster Ponds and Hatcheries, 1894–1909; memo 30 July 1894, from Assistant DM M&F, 9 August 1894, RG23, FO, vol. 243, T 3163, File 1551, Shipment of Illegally Caught Lobster to United States, 1894–1913, LAC; Order in Council 0706, 28 March 1912.

40 Canada, *Annual Report of the Department of M&F 1912–13, Fisheries* (Ottawa: C.H. Parmelee, 1913), Appendix No. 6, Quebec 231, 232; Appendix No. 3 Fishery Inspectors' Reports Nova Scotia, 123, 140.

41 From Arthur Howard, 20 March 1895, RG23, FO, vol. 288, T 3213, File 2186 , Part 1, Lobster Leases – Labrador, 1895–1905, LAC; Korneski, "Development and Diplomacy," 56–7; Arthur Howard to John Costigan (Minister M&F), 20 March 1895, RG23, FO, vol. 288, T 3213, File 2186, Part 1, Lobster Leases – Labrador, 1895–1905, LAC; *St. John's Evening Telegram,* 20 April 1895; RG2, Privy Council Office, Series A-1-a, for Order in Council, see vol. 675, Reel C-3633, LAC.

42 Notes, n.d.; memo for R.N. Venning, n.d., RG23, FO, vol. 288, T-3213, File 2186, Part 1, Lobster Leases – Labrador, 1895–1905, LAC. There may have

been some parallels made between the exclusive private leases granted on the salmon rivers on the North Shore and Gaspésie. See Castonguay, *The Government of Natural Resources*, 98–100.

43 Customs dept to F. Gourdeau, DM, 30 May 1896, RG23, FO, vol. 288, T-3213, File 2186, Part 1, Lobster Leases – Labrador, 1895–1905, LAC.

44 Canada & The South African War, 1898–1902, www.warmuseum.ca/cwm /exhibitions/boer/gathoward_e.shtml; "The Man with the Gatling Gun," *The New York Times*, 18 May 1885; Carman Miller, *Painting the Map Red: Canada and the South African War, 1899–1902* (Montreal and Kingston: McGill-Queen's University Press, 1993), 446.

45 "Brownsburg," Laurentian Heritage Webmagazine, accessed 24 May 2022, http://laurentian.quebecheritageweb.com/article/brownsburg; Thomas C Brainerd (1837–1910) Contributions to Professional Engineering, www. engineeringhistory.on.ca/index.php?id=14; "Gatling Gun Howard," *Globe*, 20 January 1900, p. 3.

46 To Minister of Fisheries from lawyer in St. John's NF, 9 September 1895, RG23, FO, vol. 288, T-3213, File 2186, Part 1, Lobster Leases – Labrador, 1895–1905, LAC.

47 "Those Canadian Seizures," *Evening Telegram* (St. John's), 5 September 1895.

48 R.N. Venning memo 1896, RG23, FO, vol. 288, T 3213, File 2186, Part 1, Lobster Leases – Labrador, 1895–1905, LAC. The first season was at least successful, as he was able to pack 1,000 cases of lobster. *Military Gazette*, 15 November 1894; *Monetary Times, Trade Review and Insurance Chronicle*, 2 October 1896, p. 461.

49 Canada, *Annual Report Department of M&F Fisheries 1901* (Ottawa: S.E. Dawson, 1902), 183; Canada *Annual Report of the Fisheries Branch 1920* (Ottawa: Thomas Mulvey, 1921), 146; "Anticosti's Future: Island the Great Lobster Producing Fishery on the Globe," *Livonia Gazette*, 1 March 1901, p. 1.

50 "Peggy's Cove Fishermen," to Department of Naval Affairs, 1 January 1912; RG23, FO, vol. 285, T 3209, File 2154, Part 7, Illegal Lobster Fishing 1910–12, LAC.

51 Jacoby, *Crimes of Nature*, 195; James C. Scott, *The Art of Not Being Governed: An Anarchist History of Upland Southeast Asia* (New Haven: Yale University Press, 2009), xii.

52 "Port L'Hebert News," *Liverpool Advance*, 10 June 1903, p. 3.

53 Memo, 4 March 1903, RG23, FO, vol. 296, T 3221, File 2314, Part 6 1901–4, Lobster Regulations, 1900–194; O.S.V. Spain, Fisheries Protective Service to DM M&F, 15 August 1894. RG23, FO, vol. 253, T 3173, File 1645 Lobster Regulations, 1894–1905, LAC.

54 Acheson, "The Lobster Fiefs"; Acheson, *Lobster Gangs*.

55 Acheson, *Lobsters Gangs*, 1, 2; Acheson, *Capturing the Commons*, 20; See also Corson, *The Secret Life of Lobsters*; Ronald P. Formisano, *The Great Lobster*

War (Amherst: University of Massachusetts Press, 1997); Croteau, *Cradled in the Waves*, 84.

56 Kevin H. Kelly, "A Summary of Maine Lobster Laws and Regulations, 1820–1992," Lobster Informational Leaflet #19, Lobster and Crab Fisheries Division, Maine Department of Marine Resources, 1992.

57 Stuart Beaton quoted in DFO, *A Brief History of the Lobster Fishery*, 28. This mirrors Acheson's "perimeter-defended ownership" once widespread in Maine. Acheson, "The Lobster Fiefs Revisited: Economic and Ecological Effects of Territoriality in the Maine Lobster Industry," in *The Question of the Commons: The Culture and Ecology of Communal Resources*, ed. Bonnie J. McCay and James M. Acheson (Tucson: University of Arizona Press, 1987), 45.

58 Wagner and Davis, "Property as a Social Relation."

59 J. Brownstein and J. Tremblay, "Traditional Property Rights and Cooperative Management in the Canadian Lobster Fishery," *Lobster Newsletter* 7, no. 1 (July 1994): 5.

60 W.H. Venning to M. Bowell Acting Minister M&F, 30 June 1883; Venning to Minister M&F, 10 July 1883; Memo for Minister M&F by W.F. Whitcher, 4 July 1883, Fishing – Disputes in Connection with Lobster Fishing Grounds in New Brunswick – From the Department of M&F 1883. RG13, Justice, Series A-2, vol. 2244, File: 1883–1223, Old Reference: 1883–53, LAC.

61 Korneski, "Development and Diplomacy," 57.

62 Flint, "Lobster Fishery of Southwest Nova Scotia," 116, citing Anthony Scott and Maurice Tugwell, *The Public Regulation of Commercial Fisheries in Canada. Case Study no. 1, The Maritime Lobster Fishery* (Ottawa: Economic Council of Canada, 1981), 26–7; DeWolf, *Lobster Fishery of the Maritime Provinces*, 18.

63 Cheverie, "Johnston's Lobster Factory, 1935–1945." We see similar themes in "Customs and Competition," The American Lobster, River John Community Access Program Committee, 2002, accessed 15 February 2024, www.parl.ns.ca/lobster/customs.htm.

64 "Sea Heritage: Oral Histories from Caribou," p. 12; Letterhead J.A. Doyle of Pugwash, "Lobster Packer and Lumber," 2 July 1937, "Factories Along the Northumberland Strait," NFM.

65 "Lobstermen Off at Dawn," *Halifax Herald*, 1 May 1939, p. 3. See also Edward MacDonald, "Running the Lines," *Island Magazine* 10 (Spring/Summer 1986): 9–12; accessed 21 July 2022, https://islandarchives.ca/islandora/object/vre:islemag-batch2-249; To Minister M&F, 20 March 1896, RG23, FO, vol. 294, Reel T-3219, File: 2314, Parts: 1 Lobster Regulations, 1895–6, LAC; "Race for Best Grounds," *CF*, June 1952, 26.

66 H.S. Kaiser to Minister M&F, 11 April 1904, RG23, FO, vol. 296, T 3222, File 2314, Part 7, Lobster Regulations, 1904–6, LAC.

67 Forwarded by Théotime Blanchard, MP Gloucester, 20 March 1896, RG23, FO, vol. 294, T 3219, File 2314, Part 1, Lobster Regulations, 1895–6; Joseph

Poirier to Minister M&F, 16 February 1905, RG23, FO, vol. 296, T 3222, File 2314, Part 7, Lobster Regulations, 1904–6, LAC.

68 Charles Stayner, Halifax, 19 September 1894, Eastern Shore report – Overseer George Rawlings at Musquodoboit Harbour; Robert Gaston Pope's Harbour; RG23, FO, vol. 253, T 3173, File 1645, Reports on Lobsters 1894, LAC; Canada, Order in Council 0776, 25 March 1914.

69 Director Allan Wargon, *Trap Thief* (Montreal: National Film Board of Canada 1957), 30 min.

70 Canada, Order in Council 1969–965; DFO, *A Brief History of the Lobster Fishery*, 22.

71 Bernard D. Smith, Port Medway, n.d. 1909 to "Dear Sir," RG23, FO, vol. 297, T 3981, File 2314, Part 9. Lobster Regulations 1909–10, LAC.

72 Stayner, 1894 Eastern Shore Report.

73 P.B. Waite, "Tupper, Sir Charles Hibbert," in *Dictionary of Canadian Biography*, vol. 15, University of Toronto/Université Laval, 2003–, accessed 19 July 2019, www.biographi.ca/en/bio/tupper_charles_hibbert_15E.html.

74 Robert Holman of Summerside, PEI to Louis Davies Minister M&F, 18 July 1896, RG23, FO, vol. 160, T 2848, File: 508, Part 1, Changes to Lobster Season – Nova Scotia 1894–7, LAC.

75 D.C. Fraser response, 2 June 1902, RG23, FO, vol. 296, T 3222, File 2314, Part 6A, Answers to Questions on Lobster Industry, 1902–12, LAC.

76 Found, Memo 1910.

77 For example, see Reports Fishery Overseers District 3 Nova Scotia 1907, RG23, FO, vol. 297, T 3981, File 2314, Part 9, Lobster Regulations 1909–10. LAC.

78 James Feeham, fishery guardian French Village, PEI to R.N. Venning, 11 February 1910, RG23, FO, vol. 297, T 3981, File 2314, Part 9, Lobster Regulations 1909–10, LAC.

79 Canada, Royal Commission on Eastern Fisheries, Evidence Gordon Romkey, President of the Cape LaHave Fishermen's Association, 2954, MG6 A, vol. 19, no. 2, Evidence vol. 18, Lunenburg, Liverpool and Lockeport, NSA; William Templeman, *The Washing of Berried Lobsters and the Enforcement of Berried Lobster Laws* Research Bulletin No 10 (St. John's: Department of Natural Resources, 1940), 6–7.

Chapter Four

1 Canada, *Canadian Lobster Commission*, 5.

2 Vera Schwach "Comments" on Bolster, *The Mortal Sea* editor Jacob Darwin Hamblin *H-Environment Roundtable Reviews*, 4, no. 3 (2014): 17, accessed 16 July 2020, https://networks.h-net.org/w-jeffrey-bolster-mortal-sea-fishing-atlantic-age-sail-roundtable-review-vol-4-no-3-2014.

3 Christine Keiner, *The Oyster Question: Scientists, Watermen and the Maryland Chesapeake Bay Since 1880* (Athens: University of Georgia Press, 2010).

4 Prince to T.B. Flint, MP, 27 July 1896, RG23, FO, vol. 160, T 2848, File 508, Part 1, Changes to Lobster Season – Nova Scotia 1894–7, LAC.

5 Patrice Dutil, *Prime Ministerial Power in Canada: Its Origins under Macdonald, Laurier and Borden* (Vancouver: UBC Press, 2017), 211.

6 *A Short Sketch of Professor Edward E. Prince, Dominion Commissioner of Fisheries, Ottawa* (Ottawa?: s.n., 1902?); Gough, *Managing Canada's Fisheries*, 105, 157. Prince removed himself from almost all department business in 1909 and focused on international issues.

7 Canada, M&F Committee, *Evidence Taken Before the M&F Committee Respecting the Lobster Industry During the Session of 1909* (Ottawa: C.H. Parmelee, 1909), 9; "Death of Former Supt of Fisheries Robt. N. Venning," *Ottawa Citizen*, 18 March 1929, p. 4.

8 Johnstone, *The Aquatic Explorers*, 71. See, for example, Canada, *Evidence*, 9, 10, 196.

9 Jennifer Hubbard, "In the Wake of Politics: The Politics and Economic Construction of Fisheries Biology, 1860–1970," *Isis* 105, no. 2 (June 2014): 364–5, 369, 364; Canada, *Annual Report of the Department of M&F 1873* (Ottawa: I.B. Taylor, 1874), Appendix O, "Special Report on Certain Petitions Against the Regulation Affecting the Lobster Fishery," 149.

10 Barbara J. Messamore, "Diplomacy or Duplicity? Lord Lisgar, John A. Macdonald, and the Treaty of Washington, 1871," *Journal of Imperial and Commonwealth History* 32, no. 2 (May 2004): 29–53.

11 Edward E. Prince, "Notes on the Natural History of the Lobster with Special Reference to the Canadian Lobster Industry," *Special Reports, Supplement No. 1 to the 29th Annual Report of the Department of M&F* (Ottawa: S.E. Dawson, 1897), xxxvi; Canada, *Annual Report of Department of M&F 1895 Fisheries* (Ottawa: S.E. Dawson, 1896), xii; Hubbard, *Science on the Scales*, 5.

12 Flint, "The Lobster Fishery of Southwest Nova Scotia," 166.

13 Prince, "Notes on the Natural History," 1.

14 Mario Mimeault, "Wakeham, William," in *Dictionary of Canadian Biography*, vol. 14, University of Toronto/Université Laval, 2003–, accessed 16 July 2020, www.biographi.ca/en/bio/wakeham_william_14E.html.

15 Canada, *Canadian Lobster Commission*, 39.

16 Wakeham, *Report*, 11.

17 Andrew Macphail, *Discoloration in Canned Lobster: Report of an Inquiry into the Causes Leading to a Deterioration in the Quality of Canned Lobsters* (Ottawa: S.E. Dawson, 1898).

18 Hubbard, *Science on the Scales*; A.P. Knight, *Official Report upon Lobster Conservation in Canada: Being the Results of Investigations Carried on Under*

the Biological Board, with the Aid of Officers Instructed by the Department of Naval Service During the Season of 1916 (Ottawa: J. de L. Taché, 1917); A.P. Knight, "Lobster Mating: A Means of Conserving the Lobster Industry," *Science* 44 (8 December 1916): 828–32; Mona Gleason, *Small Matters: Canadian Children in Sickness and Health, 1900–1940* (Montreal and Kingston: McGill-Queen's University Press, 2013); *The Ontario Public School Hygiene* (Toronto: Copp Clark, 1910); "A.P. Knight MA MD FRCS," *Queen's Review* 2 (1928): 97–9; Flint, "The Lobster Fishery of Southwest Nova Scotia," 168; Johnstone, *The Aquatic Explorers*, 58.

19 Canada, Department of Naval Service, *Annual Report of the Fisheries Branch, 1919* (Ottawa: Thomas Mulvey, 1920), 59.

20 Eric L. Mills, "Science in Canada: The Context of the Biological Board of Canada's St. Andrews Biological Station," in *Century of Maritime Science*, 42.

21 Frances Anderson, "The Demise of the Fisheries Research Board of Canada: A Case Study of Canadian Research Policy," *HSTC Bulletin Journal of the History of Canadian Science, Technology and Medicine* 8, no. 2 (27) (December 1984): 15; Hubbard, *Science on the Scales*; Johnstone, *The Aquatic Explorers*.

22 "Fish Breeding," *Annual Report Department of Fisheries 1889* (Ottawa: Brown Chamberlin), 6.

23 Robert H. Cook, "Aquaculture Research and Development at the St. Andrews Biological Station, 1908–2008," in *A Century of Maritime Science*, 360.

24 Agreement 1 July 1904, His Majesty King Edward VII and H.E. Baker, Lobster packer, Gabarus, Cape Breton, NS, RG23, FO, vol. 243, T 3162, File 1551, Part 2, Agreement with H.E. Baker for Lobster Seed Collection at Gabarus, CB, LAC.

25 L.S. Ford to Minister of M&F, 8 August 1897, RG23, FO, vol. 295, T 3220, File 2314, Part 3, 1897–8, Lobster Regulations, 1896–8, LAC.

26 Hubbard, *Science on the Scales*, 95.

27 "The Lobster and Lobster Fishery," *CF*, May 1937, p. 13. In New Brunswick, hatcheries were established at Shemogue (1903), Shippegan (at Chiasson Office on Lamèque Island, 1904), and Buctouche (1912). Prince Edward Island had two lobster hatcheries: one at Blockhouse Point near Charlottetown (1904) and another in the east at Georgetown (1909). In Quebec a temporary hatchery had operated near Gaspé in the 1890s, but two permanent hatcheries opened in 1910, at Port Daniel and House Harbour (Havre-aux-Maisons), Magdalen Islands. Nova Scotia eventually hosted eight hatcheries. In addition to Bay View, hatcheries were constructed at Canso (1905), Margaree Harbour (1911), Isaac Harbour (1911), Arichat (1911), Tracadie (1911), and Little Bras d'Or Harbour (1912). Proposed hatchery at Long Beach Pond, Digby Co., 22 January 1895, RG23, FO, vol. 270, T 3193, File 1901, Lobster Ponds and Hatcheries, 1894–1909, LAC.

28 Notes: Canadian Lobster Hatchery Visit, Bay View, Caribou Cove, Little Entrance, NS, 8 July 1893, RG22 US Fish and Wildlife Service, Office of the Commissioner of Fish and Fisheries, Container 18, National Archives; Alfred Ogden, Parliament of Canada, House of Commons, Parlinfo, accessed 17 July 2020, https://lop.parl.ca/sites/ParlInfo/default/en_CA/People/Profile?personId=14096.
29 Found to Alfred Ogden, 31 March 1914, RG23, FO, vol. 340, T 4027, File 2950, Part 1, Supplies and Repairs Charlottetown Lobster Hatchery, 1904–14; Notes: A.H. Whitman and Sons, RG23, FO, vol. 354, T 4043, File 3080, Part 1, Steamer for Canso Lobster Hatchery, Nova Scotia, 1905–13, LAC.
30 F. Gourdeau, DM M&F to Napoleon LeBlanc, 13 June 1903, RG23, FO, vol. 341, T 4028, File 2956, Part 1, Shemogue Lobster Hatchery – New Brunswick, 1903–14, LAC.
31 H.R. Emmerson to Préfontaine, Minister M&F, 29 December 1904; memo n.d.; Emmerson to Préfontaine, 10 February 1905 and 27 February 1905, Shemogue Lobster Hatchery.
32 Emmerson to Brodeur, 3 December 1907; Alfred Ogden to Prince, 1 July 1903; R.N. Venning to Napoleon LeBlanc, 10 June 1904; Report: W. Atkinson, 31 November 1904; Préfontaine to Emmerson, 14 December 1904; Emmerson to Préfontaine, 20 January 1905; Venning to LeBlanc, 22 April 1909; J.A. Roff, Supt of Fish Culture to LeBlanc, 7 June 1910; F.H. Cunningham, Supt. of Fish Culture, 24 January 1911; memo 26 February 1912, RG23, FO, vol. 343, T 4030, File 2980, Part 1; Correspondence – Shemogue Lobster Hatchery, 1903–14, LAC.
33 Memo: Cunningham, 11 August 1903; Onésiphore Turgeon to Préfontaine, 28 August 1903, RG23, FO, vol. 344, T 4031, File 2994, Parts 1, Lobster Hatchery – Caraquet, New Brunswick, 1903–04, LAC.
34 Theobald M. Burns to John Douglas Hazen, 26 February 1912, RG23, FO, vol. 347, T 4035, File 3022, Part 1, Employment – Shippegan Lobster Hatchery, 1903–14, LAC.
35 Burns to Hazen, 26 February 1912; Hazen to Burns, 30 January 1912, Employment – Shippegan Lobster Hatchery.
36 Burns to Hazen, 31 January 1912, Employment – Shippegan Lobster Hatchery.
37 Sebastien Savoy to Found, 27 February 1912; Found to Savoy, 9 March 1912; memo: 20 July 1912, Employment – Shippegan Lobster Hatchery. See also Village Historique Acadien, "Lobster Hatchery."
38 Cook, "Aquaculture Research," 360.
39 "Lobster Hatcheries Closed," CF, April 1918, p. 674.
40 To Alfred Ogden Fishery Overseer Bedford NS, 24 July 1894, RG23, FO, vol. 252, T 3173, File 1643, Lobster Eggs for British Columbia, 1894–1914, LAC.

41 Robert R. Stickney, *Aquaculture of the United States: A Historical Survey* (New York: Wiley, 1996), 60; California Commissioners of Fisheries, *Report of the Commissioners of Fisheries of the State of California for the 1879*, 14; Richard Rathburn, "The Transplanting of Lobsters to the Pacific Coast of the United States," *Bulletin of the United States Fish Commission* 8 (1888): 455, 461, 463, 468; "The Shellfish Wars: Lobsters, Shoalies, Olys, Atlantics and Pacifics Vie for Supremacy," *Chinook Observer*, 7 June 2016, accessed 17 July 2020, www.chinookobserver.com/life/the-shellfish-wars-lobsters-shoalies-olys-atlantics-and-pacifics-vie/article_a0efadbf-8785-5ae2-9672-3e194d0c4205. html; California, Report of the State Board of Fish Commissioners *Appendix to the Journal of the Senate and Assembly* 7 (1893): 10.

42 Leslie Scattergood, *The American Lobster Homarus Americanus* Fishery Leaflet 74, n.d., 2, accessed 18 July 2020, https://spo.nmfs.noaa.gov/sites /default/files/legacy-pdfs/leaflet74.pdf.

43 "Lobsters for the Pacific," *Cape Sable Advertiser*, 24 November 1887; "The Shellfish Wars."

44 Canada, *House of Commons Debates*, 6 August 1891 (Charles Tupper, Con) 7th Parliament 1st vol. 2, 3529.

45 To Alfred Ogden, Fishery Overseer Bedford NS, 24 July 1894, RG23, FO, vol. 252, T 3173, File 1643, Lobster Eggs for British Columbia, 1894–1914, LAC.

46 Ogden to Samuel Wilmot, Chief Superintendent, 14 August 1894, Lobster Eggs for British Columbia, 1894–1914.

47 Wilmot to Ogden, 3 September 1894, Lobster Eggs for British Columbia, 1894–1914. Member of Parliament (unclear signature), 22 May 1895, Lobster Eggs for British Columbia, 1894–1914. Canada, *House of Commons Debates*, 19 May 1888 (Edward G. Prior, Con) 6th, 2nd session vol. 26, 1601.

48 Memo Planting of Lobsters & Oysters in BC, n.d.; Prince to C.A. Stayner, Fishery Officer, 22 July 1896, Lobster Eggs for British Columbia, 1894–1914.

49 Report (probably Prince), 20 July 1896, Lobster Eggs for British Columbia, 1894–1914.

50 Memo (Prince's script) c. 1898, Lobster Eggs for British Columbia, 1894–1914.

51 G. Clifford Carl and C.J. Guiguet, *Alien Animals in British Columbia*, Department of Education, Handbook No. 14 (Victoria: British Columbia Provincial Museum, 1958), accessed 18 July 2020, https://ibis.geog.ubc .ca/biodiversity/efauna/AlienSpeciesinBritishColumbiaHistorical Records.html; Canada, Department of M&F, *Fisheries Annual Report, 1905* (Ottawa: S.E. Dawson, 1906), 285.

52 Davies, Minister M&F to J.J. Johnson, Western Whitstable Oysters, 27 April 1901, Lobster Eggs for British Columbia, 1894–1914.

53 Telegram Prince to G.W. Taylor, Inspector of Fisheries, 14 June 1905, Lobster Eggs for British Columbia, 1894–1914; See Parenteau, "Care, Control and Supervision."

54 Cunningham, Superintendent of Fish Culture, memo re: Lobster in Pacific, 11 July 1907, Lobster Eggs for British Columbia, 1894–1914.

55 Rodd to Williams, RS&Co.; Cunningham 1907 memo, Lobster Eggs for British Columbia, 1894–1914.

56 J.A. Rodd, Director of Fish Culture to Richard H. Williams, Manager Robin, Jones and Whitman, 12 June 1930, MG6 A, vol. 24, 2, Correspondence, Minutes, Clippings, 1913–34. RS&Co., NSA; Canada, Department of M&F, *Fisheries* 1907 (Ottawa: S.E. Dawson,1907), 303; Memo, Prince, n.d.; memo, 25 February 1909, Lobster Eggs for British Columbia, 1894–1914.

57 Victoria Board of Trade to unknown, 2 February 1909, Lobster Eggs for British Columbia, 1894–1914. G. Doutre, Purchasing and Contract Agent to Neville, 23 April 1909; Cunningham report, 23 September 1909; memo, 25 February 1909; telegram, 9 April 1909, Lobster Eggs for British Columbia, 1894–1914 and Rodd to Williams, RS&Co.

58 Taylor to Cunningham, 17 March 1909; Taylor to Cunningham, 27 March 1909; Taylor to Cunningham, 15 November 1909, Lobster Eggs for British Columbia, 1894–1914.

59 Taylor to Cunningham, 18 January 1910; Taylor to Cunningham, 15 November 1909; Taylor to Cunningham, 18 January 1910, from G.W. Taylor, Lobster Eggs for British Columbia, 1894–1914.

60 Memo, Prince, 4 April 1912, Lobster Eggs for British Columbia, 1894–1914.

61 Clipping *Pacific Fisherman*, January 1914; memo to Professor A.B. Macallum from unknown, 5 May 1914; Macallum to DM M&F, 2 June 1914, Lobster Eggs for British Columbia, 1894–1914.

62 Macallum to DM M&F, 2 June 1914, Lobster Eggs for British Columbia, 1894–1914. See *J. B. L. (1934).* "Archibald Byron Macallum. 1858–1934," accessed 18 July 2020, *Obituary Notices of Fellows of the Royal Society* 1, no. 3: 287–91; Stanton, Assistant DM to Allen Myers, Victoria, 5 December 1913, RG23, FO, vol. 408, T 3396, File 4680, Part 1, Possible Establishment – Lobster Fishery in B.C. 1913, LAC.

63 Carl and Guiguet, *Alien Animals in British Columbia*; "Lobster Experiment Going Well," CF, November 1947, p. 30; "Lobster Change Ocean and Don't Like It!" CF, November 1949, p. 42; "Lobsters Transplanted in British Columbia Again," CF, August 1966, p. 14; DeWolf, *Lobster Fishery of the Maritime Provinces*, 28.

64 R.J. Ghelardi and C.T. Shoop, *Lobster (Homarus americanus) Production in British Columbia*, Fishery Research Board of Canada (Nanaimo, BC: Pacific Biological Station, 1972); J.R. Adams, "An Interim Report on an Experiment to Introduce Atlantic Lobsters into Canadian Pacific Coast Waters," Mimeo rep. (Nanaimo: Pacific Biological Station, 1948); T.H. Butler, "Re-examination of Past Attempts to Introduce Lobsters on the British Columbia Coast," Fisheries Research Board, no. 775, 1964;

C. McLean Fraser, "Possible Lobster Planting Areas on the East Coast of Vancouver Island," *Canadian Biological Board*, 1916; R.J. Ghelardi, "Progress Report on the 1965 and 1966 Lobster Introductions at Fatty Basin, Vancouver Island, BC," *Fisheries Research Board*, 1967.

65 Nickerson to J.H. Sinclair, 3 June 1918, RG23, FO, vol. 1218, file 726-6-5, Part 1 quoted in Hubbard, *Science on the Scales*, 100.

66 "Saving the Canned Lobster Business," *CF*, July 1918, p. 834.

67 Canada, *Commission on Lobster Industry*, 25.

68 Adam Ashforth, "Reckoning Schemes of Legitimation: On Commissions of Inquiry as Power/Knowledge Forms," *Journal of Historical Sociology* 3, no. 1 (March 1990): 7; E.A. Heaman, *Tax, Order, and Good Government: A New Political History of Canada, 1867–1917* (Montreal and Kingston: McGill-Queen's University Press, 2017), 240.

69 Stephen J. Cole, "Commissioning Consent: An Investigation of the Royal Commission on the Relations of Labour and Capital, 1886–1889," (PhD diss., Queen's University, 2008); Thomas Lockwood, "A History of Royal Commissions," *Osgoode Hall Law Journal* 5, no. 2 (October 1967): 172–209; Barbara Lauriat, "'The Examination of Everything': Royal Commissions in British Legal History," *Statute Law Review* 31, no. 1 (April 2010): 24–46; Gregory J. Inwood and Carolyn M. Johns, eds. *Commissions of Inquiry and Policy Change: A Comparative Analysis* (Toronto: University of Toronto Press, 2014); Neil John Bradford, *Commissioning Ideas: Canadian National Policy Innovation in Comparative Perspective* (Toronto: Oxford University Press, 1998); R. MacGregor Dawson, rev. Norman Ward, *The Government of Canada*, 5th ed. (Toronto: University of Toronto Press, 1973), 198–9; R. Van Loon and M. Whittington, *The Canadian Political System*, 4th ed. (Toronto: McGraw-Hill Ryerson, 1987), 496–8; J.E. Hodgetts, "The Role of Royal Commissions in Canada," *Proceedings*, Institute of Public Administration in Canada, 3rd Annual Meeting (Toronto: Institute of Public Administration in Canada, 1951); R.M. Fowler, "The Role of Royal Commissions," in *Economic Policy Advising in Canada*, ed. David D. Smith (Toronto: C.D. Howe Institute, 1981); Barbara Freeman, *The Satellite Sex: The Media and Women's Issues in English Canada, 1966–1971* (Waterloo, ON: Wilfrid Laurier Press, 2001); A. Paul Pross, *Group Politics and Public Policy*, 2nd ed. (Toronto: Oxford, 1992); Jessica Squires, "Creating Hegemony: Consensus by Exclusion in the Rowell-Sirois Commission," *Studies in Political Economy* 81 (Spring 2008): 159–90.

70 Dutil, *Prime Ministerial Power*, 87–8, 101, 132.

71 Lori A. Allen, "Determining Emotions and the Burden of Proof in Investigative Commissions to Palestine," *Comparative Studies in Society and History* 59, no. 2 (2017): 385–414.

72 Order in Council 1919–02–5, 30 January 1919, Appointment Ward Fisher Commissioner to investigate whether lobster fishing or canning carried

on out of season in the Subdivision under S. Vierari now Barachois and as to efficiency of the fishery officers, RG2, Privy Council Office, Series A-1-a, LAC. For Order in Council see vol. 1216, Commission to Investigate Illegal Lobster Fishing in the Shediac Sub-Division of New Brunswick.

73 Squires, "Creating Hegemony," 159.

74 In the 1909 testimony, William Whitman, MLA for Guysborough, described himself as a "fisherman all his life," although the 1901 and 1911 census lists his occupation as a farmer, 139. Canada, M&F Committee, *Annual Report of the Department of M&F, 1909, Special Appended Report II: Evidence Taken Before the M&F Committee Respecting the Lobster Industry* (Ottawa: C.H. Parmelee, 1909).

75 "The Lobster Commissioners," *Cape Sable Advertiser*, 29 September 1887.

76 Prince to Gourdeau, 12 December 1898, RG23, FO, vol. 323, T 4008, File 2750, Part 1, 1898–9, Lobster Commission 1898–1910, LAC.

77 Notice of Maclean Commission meeting in Yarmouth, 12 and 13 December 1927, Fonds Père Clarence d'Entremont, Series Q, Scrapbook, Les Archives Père Clarence d`Entremont, Musée Acadien des Pubnicos, Lower West Pubnico.

78 *Canada Gazette*, Order in Council 1887–1467, 7 July 1887.

79 S.C. Campbell, "Hunter-Duvar, John," in *Dictionary of Canadian Biography*, vol. 12, University of Toronto/Université Laval, 2003–, accessed 20 July 2022, www.biographi.ca/en/bio/hunter_duvar_john_12E.html. With appreciation to Ed MacDonald for this point.

80 *Canada Gazette*, Order in Council 1888–0130, 1 February 1888.

81 Prince Edward Island Legislative Documents Online, Biographies of Members of the Legislative Assembly, accessed 13 November 2020, www.peildo.ca/islandora/object/leg:27536. Campbell, "Hunter-Duvar, John."

82 Alfred Ogden biography, accessed 18 July 2020, https://bdp.parl.ca/sites/ParlInfo/default/en_CA/People/Profile?personId=14096.

83 Canada, *Sessional Papers* 1891 vol. 12, no. 17b, "Report on the Fisheries Protective Service of Canada 1889," 127–39.

84 Canada, *Report on the Lobster Industry of Canada 1892*, Supplement to the 25th *Annual Report of the Department of M&F* (Ottawa: S.E. Dawson, 1893), 31.

85 *Canada Gazette*, Order in Council 1898–2311, 27 September 1898.

86 Joseph-Israël Tarte to Fielding, 26 December 1898, MG2, vol. 506, file 13, Fielding Papers, NSA.

87 L.E. Power to Davies, 4 October 1898, Fielding Papers.

88 W.L. Dawson, packer to Davies, 6 October 1899, RG23, FO, vol. 323, T 4008, File 2750, Part 2 1899–1910, LAC.

89 Louisbourg Streets Levatte's Lane, accessed 18 July 2020, www.krausehouse.ca/krause/FortressOfLouisbourgResearchWeb/Search/indy_a_m.html#LEVATTE'S%20LANE; Dr. C.W. Parker, ed., *Who's*

Who and Why, Vol. 5, 1914 (International Press Ltd., 1914); The Political Graveyard, Nova Scotia, Anglican Church Cemetery, Louisbourg, NS, accessed 20 July 2020, http://politicalgraveyard.com/geo/ZZ/NS -buried.html.

90 D.A. MacKinnon and A.B. Warburton, eds., *Past and Present of Prince Edward Island* (Charlottetown: B.F. Bowen & Co., n.d.), 444, accessed 22 December 2020, http://data2.collectionscanada.gc.ca/080027/ amicus-4651885_07.pdf; Prince Edward Island Legislative Documents Online, Biographies of Members of the Legislative Assembly, Etienne-E (Stephen) Gallant, accessed 20 July 2020, www.peildo.ca/fedora/ repository/leg:27514. Frederick William Wallace, "Scholar Poet and Fighting Fisherman: A Character Study of Moses Nickerson," CF, January 1929, pp. 18–20.

91 Canada, *Canadian Lobster Commission*, 5–6.

92 E.C. Bowers, Westport to Prince, 5 October 1898, RG23, FO, vol. 323, T 4008, File 2750, Part 1 1898–9, Lobster Commission 1898–1910, LAC.

93 Canada, *Canadian Lobster Commission*, 39.

94 Canada, *Canadian Lobster Commission*, 23, 27–8.

95 MacKinnon and Warburton, *Past and Present*, 444.

96 Canada, *Canadian Lobster Commission*, 25.

97 Canada, *Canadian Lobster Commission*, 39.

98 Canada, *Commission to Investigate Grievances and Complaints Existing in Regard to Salmon and Lobster Fisheries in Gloucester County, New Brunswick*, 1904, accessed 18 July 2020, https://epe.lac-bac.gc.ca/100/200/301 /pco-bcp/commissions-ef/prince1904-eng/prince1904-eng.pdf.

99 Canada, *Commission to Inquire into the Herring and Sardine Industry*.

100 Canada, *Commission to Inquire*, 5–6.

101 Wakeham, *Report*; *Canada Gazette*, Order in Council 1909–1470, 21 June 1909. Clippings, RG27, Labour Canada, vol. 296, Reel T-2686 File: 3135, Lobster Packers – Gabarus, Nova Scotia 1909, LAC.

102 Mimeault, "Wakeham, William."

103 Wakeham, *Report*, 3.

104 Canada, *Lobster Fishery Evidence Taken Before Commander William Wakeham, MD* (Ottawa: C.H. Parmelee, 1909), vol. 2, 988, 831.

105 RG2, Privy Council Office, Series A-1-a. For Order in Council, see volume 1042 PC 1901 4 July 1912, LAC.

106 Éloi DeGrâce, "O'Leary, Henry," in *Dictionary of Canadian Biography*, vol. 12, University of Toronto/Université Laval, 2003–, accessed 18 July 2020, www.biographi.ca/en/bio/o_leary_henry_12E.html. Prince Edward Island Legislative Documents Online, Biographies of Members of the Legislative Assembly, John McLean, accessed 18 July 2020, www.peildo. ca/fedora/repository/leg:25533. See Henry Roper, "The Halifax Board of

Control: The Failure of Municipal Reform, 1906–1919," *Acadiensis* 14, no. 2 (Spring 1985): 51, ft. 24.

107 Canada, *Department of M&F 1911–12 Fisheries* (Ottawa: C.H. Parmelee, 1912), 353.

108 Canada, *Report of the Royal Commission Investigating the Fisheries of the Maritime Provinces and the Magdalen Islands 1928* (Ottawa: F.A. Acland, 1928) (*Maclean Commission Report*).

109 *Maclean Commission Report*, 6.

110 Macmillan to Catherine Macmillan, 24 October 1927, Souris, PEI, Correspondence, 1, file 08, MG 1057, Cyrus John MacMillan Fonds, McGill University Archives.

111 Macmillan to Catherine Macmillan, 13 October 1927, *Acadia*.

112 Macmillan to Catherine Macmillan, 16 October 1927, *Acadia*; 28 October, Port Hood, NS. Emphasis in original.

113 Macmillan to Catherine Macmillan, 21 November 1927, Charlottetown, PEI.

114 Macmillan to Catherine Macmillan, 5 February 1928, Montreal.

115 *Maclean Commission Report*, 102.

116 Allen, "Determining Emotions," 410.

117 Johnstone, *The Aquatic Explorers*.

118 W.A. Found, "The Lobster Fishery in Canada," in Commission of Conservation, *Sea-Fisheries of Eastern Canada* (Ottawa: Mortimer Co., 1912), 58, quoted in Flint, "The Lobster Fishery of Southwest Nova Scotia," 125.

119 Canada, Department of the Naval Service, *Fisheries Branch 1919* (Ottawa: Thomas Mulvey, 1920), 9.

120 "Andrew Halkett's Lectures to Lobster Fishermen," *CF*, April 1920, p. 98.

121 Johnstone, *The Aquatic Explorers*, 115.

Chapter Five

1 Clipping R.J. Abbott, "The Lobster Laws," *Yarmouth Times*, 4 August 1900; "Illegal Lobster Fishing 1898–1906," RG23, FO, vol. 284, Reel T-3208, File 2154 Parts: 4=1898–1902, LAC.

2 A.P. Fitzpatrick, Chief, Protection Branch, Maritime Area, "Legal Aspects of the Lobster Fishery from Federal Department of Fisheries Point of View," pp. 3–4, RG30–3/1/ Extension Department 699–710, Fisheries, SFXUA.

3 John L. McMullan and David C. Perrier, "Lobster Poaching and the Ironies of Law Enforcement," *Law & Society Review* 36, no. 4 (December 2002): 679–718; J.L. McMullan, David C. Perrier, and Norman Okihiro, "Regulation, Illegality and Social Conflict in the Nova Scotia Lobster Fishery," *Journal of Legal Pluralism and Unofficial Law* 25, no. 33 (1993): 121–46; J.L. McMullan and David C. Perrier, "Poaching vs. the Law: the

Social Organization of Illegal Fishing," in *Crimes, Laws and Communities*, ed. John L. McMullan, Stephen Smith, and Peter D. Swan (Halifax, NS: Fernwood, 1997), 29–59; "Illegal Lobstering Threatened," *CF*, August 1922, pp. 164–5.

4   Charles C. Jenkins, "Talking About Lobsters," *Maclean's*, 1 July 1921, 20, 48.

5   Clipping, Richard Williams, supplement to 10th article draft, MG6 A, vol. 24, no. 2, Correspondence, minutes, clippings. 1913–34 RS&Co., NSA.

6   Jenkins, "Talking About Lobsters," 20.

7   "How Serious Is Lobster Poaching in the Maritimes?," 2, 8-9, RG30–3/1/ Extension Department 718–29, Fisheries, SFXUA: Gough, *Managing Canada's Fisheries*, 246. Gough refers to the death of a fisheries officer in PEI.

8   McMullan and Perrier, "Lobster Poaching"; Parenteau, "A Very Determined Opposition to the Law."

9   Memo 4 March 1903, RG23, FO, vol. 296, T-3221, file 2314, Part 5, 1900–1, Lobster Regulations 1900–4, LAC.

10   Memo 1907, RG23, FO, vol. 297, T-3222, file 2314, Part 8, Lobster Regulations 1906–9, LAC.

11   Canada, *Annual Report of Department of M&F 1894 Fisheries* (Ottawa: S.E. Dawson, 1895), Appendix 1, 12.

12   William Johnston, William G.P. Rawling, Richard H. Gimblelt, and John MacFarlane, *The Seabound Coast: The Official History of the Royal Canadian Navy, 1867–1939*, Vol 1 (Toronto: Dundurn, 2011), 20; Waite, "Tupper, Sir Charles Hibbert"; Charles H. Tupper to Governor General in Council, 13 November 1891, RG13. Justice. Series A-2. vol. 2266, File 1892–63. Old Reference: 1892–10, Regulations – Draft of proposed regulations for the Lobster Fishery – From Department of Fisheries 1891–2, LAC.

13   DM M&F to F.L. Newcombe, DM of Justice, 11 May 1894; DM M&F writes to G.T. Arnett, DM Fisheries for Nova Scotia, 11 May 1894, RG23, FO, vol. 224, Reel T-3142 File 1240, Part 1, Penalty for Fishing for Undersize Lobster, LAC.

14   K.O. Martin, "Play by the Rules or Don't Play at All: Space Division and Resource Allocation in a Rural Newfoundland Fishing Community," in *North Atlantic Maritime Cultures: Anthropological Essays on Changing Adaptations*, ed. Raoul Andersen (The Hague: Mouton Publishers 1979), 277; See, for example, Canada, *1914 Revised Statutes*, Chapter 8 – Sections 46–51 to Officers wide ranging powers.

15   Canada, *Annual Report of Department of M&F 1895 Fisheries* (Ottawa: S.E. Dawson, 1896), x, 336; Canada, *Annual Report of M&F 1904 Fisheries* (Ottawa: S.E. Dawson 1905), xxxi; Canada, *Annual Report of Department of M&F 1908–09 Fisheries* (Ottawa: Parmelee, 1909), lii.

16   W.F. Tidmarsh to A. Johnston DM M&F, 31 May 1912, RG23, FO, vol. 285, T-3209, File 2154, Part 7, Illegal Lobster Fishing 1910–12, LAC.

17 Western PEI and the Northumberland Strait had a reputation for illegal activities long before it was exposed in the 1937 Commission. For example, in 1894 O.S.V. Spain drew attention to Egmont Bay and Marie Joseph, Nova Scotia. O.S.V. Spain, Fisheries Protection Service to DM M&F, 15 August 1894, RG23, FO, vol. 253, T-3173, File 1645 Lobster Regulations, 1894–1905, LAC; David MacDonald, "The Weird and Woolly War Against Lobster Poachers," *Maclean's*, 1 October 1954, 79; Fisheries Extension Committee of St. Francis Xavier University, Brief on the Lobster Industry Submitted to Hon J. Angus MacLean, Minister of Fishery, November 1962, p. 2, RG30–3/1/ Extension Department 42, 718–29, Fisheries, SFXUA.

18 Canada, *Annual Report Department of M&F 1898 Fisheries* (Ottawa: S.E. Dawson, 1899), 107.

19 A. Hatfield to A.C. Robertson, 4 October 1911, RG23, FO, vol. 173, Reel T-2939, File 673, Part 2, 1910–13, Charter of Steamers to Protect Maritimes' Lobster Fishery, 1894–1913, LAC; "Commission Concludes Sittings on Island," *Halifax Herald*, 11 May 1937, p. 3.

20 F. Goudreau, DM M&F to *Guardian*, Charlottetown, 9 July 1898, RG23, FO, vol. 284, T-3207, File 2154 Illegal Lobster Fishing, 1898, LAC.

21 R.N. Venning to R.A. Chapman, 15 August 1894, RG23, FO, vol. 253, T-3173, File 1645; Lobster Regulations, 1894–1905; emo, n.d., RG23, FO, vol. 284, T-3207, File 2154, Illegal Lobster Fishing 1898–1906, LAC.

22 Report Beach Point PEI, 28 October 1898, RG23, FO, vol. 284, T-3207, File 2154, Illegal Lobster Fishing, 1898, LAC.

23 A.C. Bertram Inspector of Fisheries to DM M&F, 15 August 1894, RG23, FO, vol. 253, T-3173, File 1645; Lobster Regulations, 1894–1905, LAC.

24 Canada, Department of the Naval Service, *Annual Report of Fisheries Branch 1918* (Ottawa: J de L. Taché, 1920), 8, 12, 13, 15.

25 Canada, Department of M&F, *Annual Report of Fisheries Branch 1921–22* (Ottawa: F.A. Acland, 1922), 393; Canada, Department of M&F, *Annual Report of Fisheries Branch 1923–24* (Ottawa: F.A. Acland, 1924), 41.

26 "Retirement Day Recalls," Fonds Père Clarence d'Entremont, Series Q, Scrapbook, Les Archives Père Clarence d'Entremont, Musée Acadien des Pubnicos et Centre de recherche, Lower West Pubnico.

27 John Morrison (Inspector) to Found, 8 June 1912, RG23, FO, vol. 285, T-3209, File 2154, Part 7, Illegal Lobster Fishing 1910–12, LAC.

28 "Operated Isolated Cannery," *Halifax Herald*, 21 April 1937, p. 3.

29 Canada, *Annual Report of the Department of M&F, Fisheries* 1893 (Ottawa: S.E. Dawson, 1894), 250.

30 Canada, *Annual Report of the Department of M&F 1892 Part II Fisheries* (Ottawa: S.E Dawson 1893), 174; Josiah Wood, Sackville, NB, to John Costigan, 10 April 1895, RG23, FO, vol. 260, T-3182, File 1707, Illegal Lobster Packing – Prince Edward Island, 1894–5, LAC.

31 Extract report Owen, 23 August 1894, RG23, FO, vol. 253, T-3173, File 1645; Lobster Regulations, 1894–1905, LAC.

32 "Lobster Factories Seized," *Liverpool Advance*, 7 July 1886; "Terror at Shelburne. Capturing a Lobster Shop – A Striking Mid-night Scene," *Cape Sable Advertiser*, 15 July 1886.

33 Evelyn Richardson, *My Other Islands* (Toronto: Ryerson, 1960), 65.

34 MacDonald, "The Weird and Woolly War," 79.

35 Isaac Vigneault Fishery Officer (1915–88) History and Biographies Fisheries & Oceans Canada Gulf Region, Conservation and Protection, NFM.

36 MacDonald, "The Weird and Woolly War," 79; Canada, Department of Fisheries, *Annual Report 1953–54*, 13, accessed 19 July 2020, https://waves-vagues.dfo-mpo.gc.ca/Library/40768600_1953-54.pdf.

37 MacDonald, "The Weird and Woolly War," 79; "Crackdown on Illegal Lobster Fishing," *CF*, July 1954, p. 19.

38 Fisheries Extension Committee Brief 1962, p. 9; "How Serious Is Lobster Poaching?," 3.

39 A.M. Hatfield to Gourdeau, DM M&F, 11 March 1903, RG23, FO, vol. 284, T-3207, File 2154, Illegal Lobster Fishing 1898–1906, LAC.

40 "How Serious Is Lobster Poaching?," 8–9; Gough, *Managing Canada's Fisheries*, 246.

41 Canada, Department of Fisheries, *Annual Report 1954–55*, 12, accessed 19 July 2020, https://waves-vagues.dfo-mpo.gc.ca/Library/40768600_1954-55.pdf; Fisheries Extension Committee Brief 1962, p. 4; "How Serious Is Lobster Poaching?," 8.

42 MacDonald, "The Weird and Woolly War," 14.

43 Stanley Dudka, Fishery Officer (1923–2008); Isaac Vigneault, clipping *Novascotian*, 4 February 1989, History and Biographies, NFM.

44 See Parenteau, "Care, Control and Supervision." R.A. Chapman to Department, 17 September 1898; Report, 9 December 1898, RG23, FO, vol. 284, T-3207, File 2154, Illegal Lobster Fishing, 1898, LAC.

45 In 2015 a coast guard vessel named in his honour. "La petite-fille d'Agapit LeBlanc lui rend homage," https://ici.radio-canada.ca/nouvelle/730396/gloria-agapit-leblanc posted 9 July 2015, accessed 20 July 2020.

46 "MP Linked to Lobster Inquiry," *Charlottetown Guardian*, 17 February 1937, p. 5.

47 MacDonald, "The Weird and Woolly War," 14.

48 Clippings *Evening Telegram*, 22 May 1959; *Evening Telegram*, 12 May 1959; *Daily News*, 13 May 1959; *Evening Telegram*, 19 September 1958; "Armed Lobster Poachers Worry Fisheries Department," *Daily News*, 19 September 1958, GN34/2 Box 48 Department of Fisheries, Departmental Files 32/1 vol. 1 Lobster Fishing Regulations, PANL.

49 Fitzpatrick, "Legal Aspects of the Lobster Fishery," 11.

50 Goudreau to H.C. Reynolds, Green Harbor, 22 February 1904, RG23, FO, vol. 284, T-3207, File 2154, Illegal Lobster Fishing 1898–1906, LAC.

51 "Port L'Hebert News," *Liverpool Advance*, 23 December 1903.

52 Jacoby, *Crimes Against Nature*, 2; Chapman to Department, 28 October 1898: Report. Beach Point, PEI, 28 October 1898, RG23, FO, vol. 284, T-3207, File 2154, Illegal Lobster Fishing 1898; Josiah Wood, Sackville, NB, to John Costigan MP, 10 April 1895, RG23, FO. vol. 260, T-3182, File 1707, Illegal Lobster Packing PEI, 1894–5, LAC.

53 Letters to Mr. Nass, Fishery Officer 1927, 11 October 1927, Fisheries Museum of the Atlantic Archives and Library, Lunenburg. See also, Loo, *States of Nature*.

54 Personal, to W.J. Kenough, Minister of Fisheries and Co-operatives, 22 March 1952, GN34/2 M&F, Box 48 Department of Fisheries, Departmental Files 32/1 vol. 1 Lobster Fishing Regulations, The Rooms.

55 MacDonald, "The Weird and Woolly War," 14; Fisheries Extension Committee Brief 1962, p. 3.

56 Extract Report, Owen, 23 August 1894, RG23, FO, vol. 253, T-3172 File 1645, Lobster Regulations 1894–5; Fielding to Brodeur, 18 January 1908, RG23, FO, vol. 285, T-3208, File 2154, Part 6, Illegal Lobster Fishing 1906–10, LAC.

57 "Eastern Canadian Fishery Conference," *CF*, October 1921, p. 227.

58 DM M&F to Hackett, 11 September 1895, RG23, FO, vol. 294, T 3219, File 2314, Part 1, Lobster Regulations 1895–6, LAC; "Poaching Worse Lately," *Halifax Herald*, 20 April 1937, p. 3.

59 "Lobster Buyer," December 1897, RG23, FO, vol. 283, Reel T-3207, File 2154, Part 2, Illegal Fishing, 1896–8, LAC.

60 Found to Placide Chiasson, 7 October 1912; J.A. Delaney to Department, 26 September 1913, RG 23 FO, vol. 285, T-3209, File 2154, Part 8, Illegal Lobster Fishing, 1912–13; Bertram to Davies, 17 December 1897, RG23, FO, vol. 283, T-3207, file 2154, Part 2, Illegal Fishing, 1896–8, LAC.

61 Prince Memo re: Lobster Complaints Baie Verte, 20 April 1905, RG23, FO, vol. 296, T-3222, File 2314, Part 7, Lobster Regulations, 1904–6; Report Overseer W.C. Hobkirk Charlottetown, PEI; Report Overseer John W. Davison, Bedeque; Report Overseer J.M. McCormack, Souris, RG23, FO, vol. 297, T-3222, File 2314, Part 8, Lobster Regulations, 1906–9, LAC.

62 H.W. Longworth to Senator Ferguson, 15 October 1895, RG23, FO, vol. 294, T-3219, File 2314, Part 1, Lobster Regulations 1895–6, LAC.

63 "Canning in Woods with Help of Family," *Charlottetown Guardian*, 19 May 1937, p. 11; "Lobster Enquiry Opens at Alberton," *Charlottetown Guardian*, 14 April 1937, p. 1.

64 "Poaching Worse Lately," *Halifax Herald*, 20 April 1937, p. 3.

65 "Landed Poached Lobsters Near Officers," *Halifax Herald*, 17 April 1937, pp. 3, 4.

66 "Live Lobster Brought in from Newfoundland," *Halifax Herald*, 19 April 1937, p. 3.

67 "Live Lobster Brought in from Newfoundland," *Halifax Herald*, 19 April 1937, p. 4.

68 "Four Admit to Washing Berried Lobsters," *Halifax Herald*, 10 May 1937, p. 3.

69 "How Serious Is Lobster Poaching?," 2-4.

70 Fisheries Extension Committee Brief 1962, pp. 6, 11.

71 Fitzpatrick, "Legal Aspects of the Lobster Fishery," 10; Clippings, *Evening Telegram*, 22 May 1959; *Evening Telegram*, 12 May 1959; *Daily News*, 13 May 1959; *Evening Telegram*, 19 September 1958; "Armed Lobster Poachers Worry Fisheries Department," *Daily News*, 19 September 1958, GN34/2 Box 48, Department of Fisheries, Departmental Files 32/1 vol. 1 Lobster Fishing Regulations, PANL.

72 Draft report not paginated, RG23 FO, vol. 323, T-4008, File 2750, Part 1, 1898–9, Lobster Commission 1898–1910, LAC.

73 *Evidence Taken*, 1909, 217, 222; Testimony 29 March 1904, RG23, FO, vol. 349, T-4037, File 3027, Part 1, Lobster and Dogfish Commission, Gloucester County, NB 1904–9, LAC.

74 Fisheries Extension Committee Brief 1962, p. 11. See Mimeault, "Wakeham, William."

75 Found Memo re: Recommendations of Special Standing Committee of the House of Commons on M&F Regarding Lobster Fishery Regulations, 14 June 1910, RG23, FO, vol. 297, T-3222, File 2314, Part 8, Lobster Regulations 1906–9, LAC.

76 Draft 1899 Report, not paginated, not dated, RG 23, FO, vol. 323, T-4008, File 2750, Part 1, Lobster Commission 1898–1910, LAC; Canada, Parliament, *House of Commons Debates*, 14th Parl, 4th Sess, Vol. 4, 5 June 1925 (Alfred Edgar MacLean – Prince, Liberal) p. 3908; A.L. Barry testimony Maclean Commission, MG 6 A, vol. 16, no. 1, Evidence, vol. 1, p. 226 Campbellton, Gaspe Basin, Shippegan, and Caraquet, NSA.

77 Wakeham, *Report*, 9 11. See Mimeault, "William Wakeham."

78 *Canada Gazette*, vol. 52, no. 17, 26 October 1918, p. 4.

79 Testimony of A.J. Doucette, 4451, Maclean Commission, MG6 A, vol. 21, no. 2, Evidence, vol. 11 Richibucto, Buctouche, Shediac, and St. John, NSA.

80 21 April to 19 August. See Brun, *La ruée vers le homard*, 20.

81 "Lobster Fishery Probe Under Way," *Charlottetown Guardian*, 8 January 1937, p. 8; W.D. Hamilton, "Loggie, Andrew," in *Dictionary of Canadian Biography*, vol. 15 (Toronto: University of Toronto/Université Laval, 2003–), accessed 29 February 2020, www.biographi.ca/en/bio/loggie _andrew_15E.html.

82 *Report on Illegal Fishing and Canning*, 4.

83 "Illegal Fishing and Canning of Lobster on Island, Claim," *Charlottetown Guardian*, 2 February 1937, p. 1; "Four Testify at Lobster Probe," *Charlottetown Guardian*, 3 February 1937, p. 1.

84 "Illegal Fishing and Canning of Lobsters on Island Claim," *Charlottetown Guardian*, 2 February 1937, pp. 5, 14; *Canada Gazette* 70, no. 36, 6 March 1937 noting PC-429, 3 March 1937; *Report on Illegal Fishing and Canning*.

85 Captain Abraham Hatfield, Arcadia, NS, to T.B. Flint, 21 July 1902, RG23, FO, vol. 296, T-3222, File 2314, Part 6A, Answers to Questions on Lobster Industry, 1902–12, LAC.

86 Hatfield to Prince, 29 December 1904, RG23, FO, vol. 296, T-3222, File 2314, Part 7, Lobster Regulations, 1904–6, LAC; *Report on Illegal Fishing and Canning*, 4.

87 *Canada Gazette*, 2 December 1939, p. 13, re: PC 3748, 22 November 1939.

88 Joseph Poirier, Packer, Grand Anse, NB to Minister M&F, 16 February 1905, RG23, FO, vol. 296, Reel T-3222, File 2314, Part 7, Lobster Regulations, 1904–6; "Peggy's Cove Fishermen," to Department of Naval Affairs, 1 January 1912; RG23, FO, vol. 285, T-3209, File 2154, Part 7, Illegal Lobster Fishing, 1910–12, LAC.

89 "Eastern Shore Fishermen Are Pleased with Subsidized Lobster Service," *Herald*, 7 May 1930, p. 15; "Co-operative Marketing of Lobsters Proposed to Prevent Future Glut," *Herald*, 15 May 1930, p. 13.

90 Pictou Survey Committee by the Departments of Social Service and Evangelism of the Presbyterian and Methodist Churches, *Pictou District: Report on a Limited Survey of Both Rural and Urban Conditions, 1915*, 13. United Church Archives, Toronto.

91 Stephen Leahey, *Stories from the Lobster Fishery of Cumberland's Northern Shore* (Lockeport: North Cumberland Historical Society and Community Books of Lockeport, 2005), 20.

92 Sunrise Trail Museum, "1820–1920: A Century of Industry on the North Shore of Nova Scotia: Wallace to River John," Community Memories, accessed 20 July 2020, www.virtualmuseum.ca/sgc-cms/histoires_de _chez_nous-community_stories/pm_v2.php?id=story_line&lg=Englis h&fl=0&ex=00000720&sl=6700&pos=1&pf=1#19; *Rex v. Percy Langille*, 7 September 1939, Case 3316, RG39 C Cumberland Co Supreme Court, NSA; DFO, *A Brief History of the Lobster Fishery*, 17.

93 Leahey, *Stories from the Lobster Fishery*, 19; DFO, *A Brief History of the Lobster Fishery*, 17.

94 Leahey, *Stories from the Lobster Fishery*, 14, 39; "Fisherman Target for Rifle Shot," *Amherst Daily News*, 29 May 1933, p. 6.

95 "Lobster Price Is Way Down," *Oxford Journal*, 21 April 1932, p. 1; "Peace Reigns on Malagash-Wallace Front," *Halifax Herald*, 31 May 1933, p. 3.

96 Leahey, *Stories from the Lobster Fishery*, 10, 14, 20.

97 "Trouble Among Lobster Fishermen at Malagash," *Oxford Journal*, 1 June 1933, p. 1; "Want Changes Made in Lobster Laws: Fisherman Target for Rifle Shot," *Amherst Daily News*, 29 May 1933, pp. 1, 6.

98 "Shots Are Fired as Lobster Fishermen Wage War," *Halifax Chronicle*, 29 May 1933, p. 1.

99 Leahey, *Stories from the Lobster Fishery*, 39.

100 "Trouble Among Lobster Fishermen at Malagash," *Oxford Journal*, 1 June 1933, p. 1; "Fisherman Target for Rifle Shot," *Amherst Daily News*, 29 May 1933, p. 6; "Peace Reigns on Malagash-Wallace Front," *Halifax Chronicle*, 31 May 1933, p. 1.

101 "Fisherman Target for Rifle Shot," *Amherst Daily News*, 29 May 1933, p. 6.

102 "Steamer Arleux Halts Fish War," *Amherst Daily News*, 1 June 1933, p. 1.

103 "All Quiet on Lobster Front at Present," *Halifax Herald*, 30 May 1933, pp. 1, 3; "Shots Are Fired as Lobster Fishermen Wage War," *Halifax Chronicle*, 29 May 1933, pp. 1, 2.

104 "Trouble Among Lobster Fishermen at Malagash," *Oxford Journal*, 1 June 1933, p. 1; "Fisherman Target for Rifle Shot," *Amherst Daily News*, 29 May 1933, p. 6.

105 "All Quiet on Lobster Front," *Halifax Herald*, 1 June 1933, p. 3.

106 Leahey, *Stories from the Lobster Fishery*, 10, 12, 13.

107 George Tuttle King b Pugwash 1876–1941 m Amy Kennedy 1883–1939; Canada, Census, 1911, Nova Scotia, Cumberland, River Philip poll district no. 27, p. 5; obituary, *Chronicle Herald*, 30 October 1973; *Halifax Chronicle*, 17 May 1939, p. 3.

108 Leahey, *Stories from the Lobster Fishery*, 14, 19, 88.

109 "Malagash Lobster Packer Goes on Trial," *Halifax Chronicle*, 17 May 1939, p. 3.

110 "1820–1920: A Century of Industry," Canneries section, accessed 20 July 2020, www.museevirtuel-virtualmuseum.ca/sgc-cms/histoires_de_chez _nous-community_memories/pm_v2.php?id=story_line_child&fl=&lg =English&ex=00000720&sl=6700&pos=1&scpos=1.

111 Cumberland Co Supreme Court, *Rex v. Percy Langille*.

112 Cumberland Co Supreme Court, *Rex v. Percy Langille*.

113 "Admits Setting Five Buildings on Fire," *Halifax Chronicle*, 22 March 1939, p. 3.

114 Leahey, *Stories from the Lobster Fishery*, 10–12, 18, 19, 24, 54, 27.

115 MacDonald, "The Weird and Woolly War," 79; Press release, 27 October 1955, GN34/2 Box 48 Department of Fisheries, Departmental Files 32/1 vol. 1 Lobster Fishing Regulations, PANL.

116 W.F. Tidmarsh testimony and Brief Canned Fish Section of Canadian Manufacturers' Association, 1 December 1927, p. 2522, Maclean Commission, MG6 A, vol. 19, no. 1, Evidence, vol. 7, Halifax, NSA.

117 Moses Cody to Found, 6 December 1928, RG30–3/1/ Extension Department, 3896–07, Fisheries, SFXUA.

118 "Your Pot of Gold Can Be a Lobster Pot," *Antigonish Casket*, 10 May 1956, p. 5.

119 1944 lobster licence for fisherman at Blue Rocks, Fisheries Museum of the Atlantic Archives and Library, Lunenburg.

120 "Homarus Americanus to the Biologist – but Good Times and Good Living to You" *Antigonish Casket*, 7 June 1956, p. 3.

121 "Lobsters Are Everybody's Business," *Antigonish Casket*, 3 May 1956, p. 3.

122 Fisheries Extension Committee Brief 1962, p. 10.

123 See David Meren, "Lessons Learned: Settler Colonialism, Development and the UN Regional Training Centre in Vancouver, 1959–62," *BC Studies* 208 (Winter 2020/21): 65 for parallel example on the use of film.

124 "Trap Thief," *Trade News*, March 1957, p. 12.

Chapter Six

1 Jenkins, "Talking About Lobsters," 46.

2 M.C. Urquhart, "New Estimates of Gross National Product, 1870–1926. Some Implications for Canadian Development." In *Long Term Factors in American Economic Growth*, ed. Stanley L. Engerman and Robert E Gallman (Chicago: University of Chicago Press, 1986), 20–5.

3 I identify the 1896 Lobster Ridings as Bonaventure, Gaspé, Charlotte, Gloucester, Kent, Northumberland, Restigouche, Westmorland, East Prince, East Queen's, King's, West Prince, West Queen's, Antigonish, Cape Breton (two seats), Colchester, Cumberland, Digby, Guysborough, Halifax (two seats) Inverness, Lunenburg, Pictou (two seats), Richmond, Shelburne and Queen's, Victoria, and Yarmouth for a total of thirty seats and 14 per cent of the total. In 1966 the number of Lobster Ridings had dropped to twenty seats, or 7.5 per cent of the seats in the House of Commons. (Bonaventure, Gaspé, Burin-Burgeo, Humber-St. George's-Ste. Barbe, Cardigan, Egmont, Hillsborough, Malpeque, Westmorland-Kent, Restigouche, Northumberland-Miramichi, Gloucester, Carleton-Charlotte, Cape Breton Highlands-Canso, Cape Breton-East Richmond, Central Nova, Cumberland-Colchester North, South Shore, South-Western Nova, and Cape Breton-The Sydneys).

4 Five seats were determined by acclamation (the Halifax explosion settled two, and three were decided by the parties). Altogether, of the 165,427 votes cast for Unionist or Opposition candidates in the lobster ridings, only 82,379 went for the nationally successful coalition.

5 Jenkins, "Talking About Lobsters," 20, 46.

6 "Urges Change in Fisheries Act Set-Up," *Charlottetown Guardian*, 8 March 1937, p. 1.

7 Jenkins, "Talking About Lobsters," 46.
8 "Commission Concludes Sittings on Island," *Halifax Herald*, 11 May 1937, p. 3; Gordon McKean, "Lobstermen in Million Dollar Race," *Halifax Herald*, 3 May 1937, p. 8; Clipping, R.H. Williams, *Morning Chronicle*, 8 August 1918, MG6 A, vol. 30, Scrapbook of articles, newspaper clippings, etc. on the lobster industry, RS&Co., NSA.
9 Nova Scotia Provincial Elections 1867–2016, Shelburne 1911, R. Ward Fisher, Accessed 21 July 2020, https://electionsnovascotia.ca/sites/default/files/Elections1867to2017.pdf.
10 W.F. Tidmarsh to R.N. Venning, 8 May 1909, RG23, FO, vol. 297, T-3981, File 2314, Part 9, Lobster Regulations 1909–10, LAC. Mather, "W.F. Tidmarsh and Lobster Canning in Prince Edward Island."
11 Clipping, R.H. Williams, *Morning Chronicle*, 8 August 1918, MG6 A vol. 30, Scrapbook of articles, newspaper clippings, etc. on the lobster industry, RS&Co., NSA.
12 Testimony of Ernest Mark, 385–96, Maclean Commission, MG6 A, vol. 16, no. 1, Evidence vol. 1, Campbellton, Gaspe Basin, Shippegan and Caraquet, Richibucto, Buctouche, Shediac, St. John, NSA.
13 M.J. Neville to Wm. Roche MP, 5 February 1917, RG23, FO, vol. 284, T-3207, File 2154, Illegal Lobster Fishing 1898–1906, LAC.
14 Henry Le Vatte, Fishery Officer Louisbourg to Found, 10 March 1911, RG23, FO, vol. 298, T-3981, File 2314, Part 10A, Lobster Regulations – Minister's File, 19, LAC.
15 Col. A.L. Barry, Department of Fisheries to Moses Coady, 26 April 1947; Coady to Barry, 1 May 1947, RG30–3/1/ Extension Department, 3716–24, Fisheries, SFXUA.
16 F. Gourdeau, DM, M&F to E.A. Macdonald MP, 18 October 1901, RG23, FO, vol. 296, T 3221, File 2314, Part 5 1900–01, Lobster Regulations, 1900–4, LAC.
17 R.N. Venning to A.M. Hatfield, 31 January 1898, RG23, FO, vol. 295, T-3220, File 2314, Part 3 1897–8, Lobster Regulations, 1896–8, LAC.
18 Carman Miller, *A Knight in Politics: A Biography of Sir Frederick Borden* (Montreal and Kingston: McGill-Queen's University Press, 2010).
19 Venning to E.B. Eaton, Fishing Overseer, Canning, NS, 25 June 1907, RG23, FO, vol. 243, T-3163, File 1551, Shipment of Illegally Caught Lobster to United States, 1894–1913, LAC.
20 Found to Venning, 15 August 1906, RG23, FO, vol. 284, T-3207, File 2154, Illegal Lobster Fishing 1898–1906, LAC.
21 Memo, 22 April 1906, RG23, FO, vol. 285, T-3208, File 2154. Part 6 Illegal Lobster Fishing, 1906–10, LAC.
22 Memo, 10 October 1903 Re: Petition of John H. Charlton for Remission of Penalty for Selling Illegal Lobsters, RG23, FO, vol. 284, T-3207, File 2154, Illegal Lobster Fishing, 1898–1906, LAC.

23 "Lobster Inquiry Adjourns until Early in April," *Charlottetown Guardian*, 20 March 1937, p. 1.

24 Davidson to Found, 26 May 1914, RG23, FO, vol. 286, T-3210, File 2154, Part 9, Illegal Lobster Fishing, 1913–14, LAC.

25 DM Fisheries to Robert Sedgewick, DM Justice, 13 June 1891, RG13, Justice, Series A-2, vol. 81, File 1891–689, Department of Fisheries – Fine for illegal lobster fishing imposed on the Wood Harbor Packing Co., 1891, LAC.

26 Liberal NS MPs to Davies, 16 September 1896, RG23, FO, vol. 160, T-2848, File 508, Part 1, Changes to Lobster Season – Nova Scotia 1894–7, LAC.

27 Fielding memo and f 13 enclosed letter, L.E. Power to Davies, 4 October 1898, MG2 509, Folder 30 506, Fielding Papers, NSA; Carman Miller, "Fielding, William Stevens," in *Dictionary of Canadian Biography*, vol. 15 (Toronto: University of Toronto/Université Laval, 2003–), accessed 11 May 2020, www.biographi.ca/en/bio/fielding_william_stevens_15E.html.

28 Fielding to Davies, 8 July 1900; Gourdeau to Prince, 9 July 1901, RG23, FO, vol. 173, T-2939, File 673, Part 1, 1894–1909, Charter of Steamers to Protect Maritimes' Lobster Fishery, 1894–1913, LAC.

29 A.W. Swimm to Max Lane, 16 May 1964, GN34/2, Box 10 8/387, Department of Fisheries, Departmental Files, Maritime Lobsters Ltd (A.W. Swimm), PANL.

30 Jacquard, *Lobstering*, 271.

31 Préfontaine to Angus McLennan MP (Inverness), 27 June 1903, RG23, FO, vol. 213, T-3130, File 1133, Part 3, 1898–1903, Extension – Lobster Fishing Season 1898–1905, LAC.

32 Cunningham memo for the DM, 12 December 1903, RG23, FO, vol. 270, T-3193, File 1901, Lobster Ponds and Hatcheries, 1894–1909, LAC.

33 To Brodeur, 3 December 1910, RG23, FO, vol. 298, T-3981, File 2314, Part 10A, Lobster Regulations – Minister's File, 1912, LAC.

34 James Smith, Canso to Davies, 26 August 1896, RG23, FO, vol. 160, T-2848, File 508, Part 1, Changes to Lobster Season – Nova Scotia, 1894–7, LAC.

35 Georges Arsenault, "'Le Senteur': Joseph Octave Arsenault," *Island Magazine* 33 (1993): 25–9; Prince Edward Island Legislative Documents Online, www.peildo.ca/islandora/object/leg:biography.

36 "The Prowse's," Murray Harbour-Murray River, Island Narratives Program, accessed 21 July 2020, http://vre2.upei.ca/cap/node/615.

37 Clipping, *Daily Patriot*, 20 April 1891, Research PEI Lobster – Processing #6 Rider Fonds, 2009-H0014, CMHA.

38 Georges Arsenault, "Arsenault, Joseph-Octave (1828–97)," in *Dictionary of Canadian Biography*, vol. 12 (Toronto: University of Toronto/Université Laval, 2003–), accessed 26 May 2020, www.biographi.ca/en/bio/arsenault_joseph_octave_1828_97_12E.html; Frederick L. Driscoll,

"Farquharson, Donald," in *Dictionary of Canadian Biography*, vol. 13 (Toronto: University of Toronto/Université Laval, 2003–), accessed 26 May 2020, www.biographi.ca/en/bio/farquharson_donald_13E.html.

39 Théotime Blanchard, MP (Gloucester) to Tupper, 7 July 1894, RG23, FO, vol. 212, T-3129, File 1133, Part 1, Extension – Lobster Fishing Season 1894–6, LAC.

40 Jean-Roch Cyr, "Young, Robert," in *Dictionary of Canadian Biography*, vol. 13 (Toronto: University of Toronto/Université Laval, 2003–), accessed 28 May 2020, www.biographi.ca/en/bio/young_robert_13E.html; William G. Godfrey, "Black, Joseph Laurence," in *Dictionary of Canadian Biography*, vol. 13 (Toronto: University of Toronto/Université Laval, 2003–), accessed 25 May 2020, www.biographi.ca/en/bio/black_joseph_laurence_13E. html; Lewis LeBlanc, "Melanson, Olivier-Maximin," in *Dictionary of Canadian Biography*, vol. 15 (Toronto: University of Toronto/Université Laval, 2003–), accessed 25 May 2020, www.biographi.ca/en/bio /melanson_olivier_maximin_15E.html; Brun, *La ruée vers le homard*, 30, 40.

41 DeGrâce, "O'Leary, Henry."

42 Barry Cahill, "O'Leary, Richard," in *Dictionary of Canadian Biography*, vol. 16 (Toronto: University of Toronto/Université Laval, 2003–), accessed 25 May 2020, www.biographi.ca/en/bio/o_leary_richard_16E.html.

43 W.A. Spray, "Snowball, Jabez Bunting," in *Dictionary of Canadian Biography*, vol. 13 (Toronto: University of Toronto/Université Laval, 2003–), accessed 26 May 2020, www.biographi.ca/en/bio /snowball_jabez_bunting_13E.html. Dictionary of Miramichi Biography, https://archives.gnb.ca/Search/Hamilton/DMB/SearchResults. aspx?culture=en-CA&action=0&page=945.

44 James A. Fraser, *A History of the W.S. Loggie Co Ltd, 1873–1973* (Fredericton: Provincial Archives of New Brunswick, 1973); Dictionary of Miramichi Biography, Provincial Archives of New Brunswick, accessed 21 July 2020, https://archives.gnb.ca/Search/Hamilton/DMB /SearchResults.aspx?culture=en-CA&action=3&page=577&text=j5yfR8M OaV2pRDi/JJJofYfgUiMQquzIXj+WXyayfEM=.

45 Clipping, "A Notable Man of Many Interests," *The Busy East*, n.d., pp. 30–2, MG4, vol. 7, file 14, no. 25, North Shore Archive Society, Tatamagouche Heritage Centre. "The Man Behind Mephisto: Fred Magee – An Extraordinary Entrepreneur," accessed 21 July 2020, https:// tantramarheritage.ca/2002/06/the-man-behind-mephisto-fred-magee -an-extraordinary-entrepreneur/; clipping, "A Look into the Lobster Industry's Past."

46 Canada *Annual Report of the Department of M&F for the Year Ending, 30 June 1872* (Ottawa: I.B. Taylor, 1873), 30; *Annual Report of the Department of M&Fs Being for the Fiscal Year End 30ᵗʰ June 1874* (Ottawa: MacLean, Roger & Co, 1875), 19.

47 Canada, *Report of the Commissioner of Fisheries for the Year Ending 31 December 1877* (Ottawa: Maclean, Roger & Co, 1878), 158; Canada, *Department of Fisheries, Fisheries Statements for 1890* (Ottawa: Brown Chamberlin, 1890), 27; A.R. Gordon, "Lobster Fishing Industry," Canada, *Sessional Paper* 1891, vol. 8 appendix 8, 127–39.

48 Canada, *Canadian Lobster Commission*, 38–9.

49 Fielding to A.G. Blair, 12 May 1902; Fielding to Blair, 6 June 1902; Donald Farquharson; Kendall, June 1902, RG23, FO, vol. 296, T-3222, File 2314, Part 6A, Answers to Questions on Lobster Industry, 1902–12, LAC.

50 Clipping, *Daily Patriot* signed Ephraim A. Bell, Cape Traverse, 20 August 1894, RG23, FO, vol. 253, T-3173, File 1645, Part 2, Illegal Lobster Fishing PEI, 1894.

51 "How Serious is Lobster Poaching," 6–7.

52 "Eastern Canadian Fishery Conference," *CF*, October 1921, p. 227.

53 "Urges Changes in Fisheries Act Set- Up," *Charlottetown Guardian*, 8 March 1937, p. 1.

54 Circular by Prince, 12 June 1896, RG23, FO, vol. 295, T-3220, File 2314, Part 2, Lobster Regulations, 1896–8, LAC.

55 Extract James H. Whitman, Salmon River, 22 March 1894, RG23, FO, vol. 177, T-2943, File 691, Part 1, Lobster Regulations – Amendments to Fisheries Act re Size of Catch 1893–5, LAC.

56 Fielding to Préfontaine, 25 March 1904; Prince memo, 6 April 1904, RG23, FO, vol. 213, T-3130, File 1133, Part 4, 1903–5, Extension-Lobster Fishing Season 1898–1905, LAC.

57 R.E. Armstrong to Préfontaine, 17 March 1904, RG23, FO, vol. 296, T-3222 File 2314, Part 6, 1901–4, Lobster Regulations, 1900–4, LAC.

58 Clipping *Yarmouth Herald*, 13 September 1898, RG23, FO, vol. 295, T-3220, File 2314, Part 4, Lobster Regulations, 1898–1900, LAC.

59 Petition, 26 June 1894, RG23, FO, vol. 212, T-3129, File 1133, Part 1, Extension – Lobster Fishing Season 1894–6, LAC.

60 Tancook Petition, March 1910, RG23, FO, vol. 297, T-3981, File 2314, Part 9, Lobster Regulations, 1909–10, LAC.

61 Angus L. Frelick, Western Head, Queens County letter to editor *Halifax Herald* (c. 1910). MG6 A, vol. 26, no. 2, Scrapbook of newspaper clippings etc. on the lobster industry. RS&Co., NSA; West Berlin Petition, 6 November 1911; Port Joli Petition, 5 November 1911; Port Medway Petition, 5 November 1911, RG23, FO, vol. 298, T-3981, File 2314, Part 10A, Lobster Regulations – Minister's File, 1912. LAC.

62 Payne, "Local Economic Stewards"; Payne, "Becoming a Dependent Class."

63 "Lobster Packers' Case Presented by Williams," *CF*, September 1926, p. 270.

64 "Lobster Packers' Case."

65 R.H. Williams, "A Business Without Ethics," speech to Halifax Rotary Club in June 1927 *CF*, July 1927, p. 214.

66 M. Neville to Brodeur, 4 February 1907, RG23, FO, vol. 297, T-3222, File 2314, Part 8, Lobster Regulations, 1906–9, LAC.

67 W.F. Tidmarsh to Brodeur, 18 March 1909, RG23, FO, vol. 297, T-3981, File 2314, Part 9, Lobster Regulations, 1909–10, LAC; "Yesterday in the House of Assembly," *Charlottetown Guardian*, 18 March 1909, p. 1; "Budget Debate Finally Closed," 19 March 1909.

68 Ledger 1886–7, 1886 season, MG200, Henry T. d'Entremont Fonds, Series G Sub 8, Amiro Bros & d'Entremont Fish Co., Argyle Archives Lower East Pubnico.

69 Prince memo, 21 January 1904 re: Lobster meeting in Pugwash 10 December 1903, RG23, FO, vol 296, T-3221, File 2314, Part 5, 1900–1 Lobster Regulations 1900–4, LAC.

70 St. Thomas and Richibucto were the exception. List of 1908 cannery licences, n.d. RG23, FO, vol. 297, T-3981, File 2314, Part 9, Lobster Regulations, 1909–10, LAC.

71 R.N. Venning, "The M&F Committee, and the Lobster Industry," 78. Canada, Department of M&F, *1908–9 Fisheries* (Ottawa C.H. Parmelee, 1909), Special Appended Report II.

72 Colin McKay, "The Nova Scotia Fishermen's Union" (originally printed *Montreal Herald*, 31 August 1905) in Ian McKay, ed., *For a Working-Class Culture in Canada: A Selection of Colin McKay's Writings on Sociology and Political Economy, 1897–1938* (St. John's: Canadian Committee on Labour History, 1996), 233; Payne, "Local Economic Stewards"; W. Jeffrey Bolster, "Putting the Ocean in Atlantic History: Maritime Communities and Marine Ecology in the Northwest Atlantic, 1500–1800, *American Historical Review 113 (February 2008): 19–47.

73 There were reportedly 2,689 lobstermen in Maine in 1902. Maine, *Sea and Shore Fisheries 1901–1902* (Augusta, ME: Kennebec Journal, 1903), 25. Numbers in Canada are difficult to determine, as weight of lobster and number of traps and canneries were collected but not the number of men fishing. In 1904, a reported 13,981 people were employed in lobster canneries and 34,647 were fishing from inshore boats (i.e., not vessels) in New Brunswick, Nova Scotia, and Prince Edward Island. This would include all fisheries, but lobster was the most important. Lobstermen of Quebec are missing. In the first year lobster licences were introduced, 12,573 were issued. This number likely reflected lower employment in the final year of the First World War. It is unclear whether helpers held licences or how many fishermen ignored the new regulation. "Statement of the Lobster Industry in Canada during the season of 1904," *Annual Report M&F 1905* (Ottawa: S.E. Dawson, 1906), xii; 92, 122, 136; Canada, *Annual Report of Fisheries Branch 1919* (Ottawa: Thomas Mulvey, 1920), 65.

74 Robert H. Babcock, *Gompers in Canada: A Study of American Continentalism Before the First World War* (Toronto: University of Toronto Press, 1974), 119; Charles A. Scontras, "Maine Lobstermen and the Labor Movement: The Lobster Fishermen's International Protective Association, 1907," *Maine Historical Society Quarterly* 29, no. 1 (Summer 1989): 35.

75 Babcock, *Gompers*, 120.

76 L. Gene Barrett, "Underdevelopment and Social Movements in the Nova Scotia Fishing Industry to 1938," in *Underdevelopment and Social Movements in Atlantic Canada*, ed. Robert J. Brym and R. James Sacouman (Toronto: New Hogtown Press, 1979), 127–60; Judd, "Grassroots Conservation in Eastern Coastal Maine."

77 Canada, M&F Committee, *Annual Report of the Department of M&F, 1909, Special Appended Report II: Evidence Taken Before the M&F Committee Respecting the Lobster Industry* (hereafter *Evidence Taken*), 127–8; 249.

78 C.R. Fay and H.A. Innis, "The Maritime Provinces," in *The Cambridge History of the British Empire*, vol. VI: Canada and Newfoundland, ed. J. Holland Rose, A.P. Newton, and E.A. Benians (Cambridge: University Press, 1930), 660.

79 Memo to F.A. Acland, DM Labour, 17 May 1909, RG27, Labour Canada, vol. 296, T-2686 File 3135; Lobster Packers – Gabarus, Nova Scotia 1909, LAC.

80 G.J. Desbarats to Donald Scott, Fishermen's Union, Main-à-Dieu, 30 April 1909, RG23, FO, vol. 297, T-3981, File 2314, Part 9, Lobster Regulations 1909–10; Andrew Halkett, Report of Lobster Investigations, 1 December 1909, RG23, FO, vol. 323, T-4008, File 2750, Part 2, 1899–1910, Lobster Commission, 1898–1910.

81 Brodeur to A.W. Chisholm, 24 April 1909, RG23, FO, vol. 297, T-3981, File 2314, Part 9, Lobster Regulations 1909–10, LAC.

82 From Charles E. Kenny, Fishermen's Union of NS, Station No 3, n.d. c1905, RG23, FO, vol. 270, T-3193, File 1901; Lobster Ponds and Hatcheries, 1894–1909, LAC.

83 From Little Harbor Shelburne Station 2 Fishermen's Union, 23 December 1905, RG23, FO, vol. 296, T-3222, File 2314, Part 7, Lobster Regulations, 1904–6, LAC.

84 From Port Mouton Fishermen's Union of Nova Scotia Station, no. 5, 24 December 1905, Lobster Regulations, 1904–6.

85 Little Harbor Fishermen's Union of Nova Scotia, Station 2, 24 February 1912, RG23, FO, vol. 298, T-3981, File 2314, Part 10A, Lobster Regulations – Minister's File, 1912, LAC.

86 *Evidence Taken*, 128; "List of 1908 cannery licenses, Lobster Regulations 1909–1910." For example, in 1926 the Fishermen's Co-operative at Tignish forwarded 400 cases to W.F. Tidmarsh, manager for the Portland Packing

Company. See: Minutes 10 June 1926, Acc 3772, Tignish Fisheries Fonds, PARO.

87 https://royalstarfoods.com/about-us/history/; Tignish Fisheries fonds: [1925–71]; "Co-Operative Canneries in the Maritimes," *Charlottetown Guardian*, 22 June 1933, p. 2.

88 Testimony of Henry Doyle, 614, Maclean Commission, MG6 A, vol. 16, no. 2, evidence vol. 2, Summerside and Charlottetown, NSA.

89 Minutes 10 June 1926; minutes 10 February 1931, Tignish Fisheries Fonds.

90 Mary Arnold to Jacob Baker, President, United Federal Workers of America, 29 September 1940, A-122, Mary Ellicott Arnold Papers, Radcliffe Institute, Schlesinger Library, Cambridge, MA; William Daniel MacInnes, "Clerics, Fishermen, Farmers and Workers: The Antigonish Movement and Identity in Eastern Nova Scotia, 1928–1939," (PhD diss., McMaster University, 1978), 414, n. 25; Peter Ludlow, "Saints and Sinners: Popular Myth and the Study of Personalities of the Antigonish Movement," *Acadiensis* 42, no. 1 (Winter/Spring 2013): 99–126.

91 That the Dominion Department of Fisheries was a major funder of Extension Department between 1934 and 1939 is noted in MacInnes, "Clerics, Fishermen, Farmers and Workers," 333, n. 35; Ludlow, "Saints and Sinners," 118; Santo Dodaro and Leonard Pluta. *The Big Picture: The Antigonish Movement of Eastern Nova Scotia* (Montreal and Kingston: McGill-Queen's University Press, 2012). Calhoun, *Word to Say*.

92 "Lobster in Shell vs Canned," *CF*, March 1929, p. 37; "Lobster Situation in January," February 1929, p. 21.

93 MacInnes, "Clerics, Fishermen, Farmers and Workers," 363, quoting *Extension Bulletin*, 5 October 1937, p. 5; "A Story from Guysboro," in L. Brooks, ed., *A Tour of Nova Scotia Co-operatives* (Antigonish: Extension Department, 1938), 39; Peggy Feltmate, "Father Jimmy, Billy Tom and the Antigonish Movement," *Early Canadian Life* 4, no. 9 (September 1980), accessed 23 July 2020, https://peggyfeltmate.com/genealogy -social-history/articles/father-jimmy/

94 Will R. Bird, "The Story of the Remarkable Success of the Co-operative Movement in Eastern Nova Scotia," *Maclean's*, 1 August 1936, 9.

95 Dodaro and Pluta, *Big Picture*, 110; "Lobster Fishermen Threaten to 'Strike,'" *Halifax Herald*, 10 May 1930, p. 2.

96 MacInnes, "Clerics, Fishermen, Farmers and Workers," 222, 412, n. 15 and 16.

97 J.T. Croteau, *Cradled in the Waves*, 133.

98 MacInnes, "Clerics, Fishermen, Farmers and Workers," 395, 414, n.25.

99 Calhoun, *Word to Say*, 22.

100 MacInnes, "Clerics, Fishermen, Farmers and Workers," 334, n. 33 and 399, n. 3, 4. For interwar Acadian nationalism, see Nancy Carvell, "A People

Apart: New Brunswick Acadians, Conscription, and the Second World War," (MA thesis, Carleton University, 2019), 43–6.

101 Calhoun, *Word to Say*, 22.

102 To J. Angus MacLean, 6 March 1961, GN34/2 Box 48 32/2 Department of Fisheries, Departmental Files, Lobster Fishery Operations – NW Coast, PANL.

103 Appendix B-5, "Submission of Prince Edward Island Fishermen's Association," *Report of the Prince Edward Island Fisheries Development Committee* (Charlottetown, 1952), 157–8.

104 Advertisement "Attention Fishermen," *Charlottetown Guardian*, 11 April 1953, 10: "PEI Fishermen's Ass'n Area Meeting Yesterday," 4 November 1953, p. 8. In the 1950s in Maine, a more radical organization was formed in the Maine Lobstermen's Association. See Ron Formisano, *The Great Lobster War* (Amherst: University of Massachusetts Press, 1997).

105 "Lobstermen Demand 'Get Tough,'" *CF*, January 1954, pp. 24–5.

Chapter Seven

1 *Canada Year Book* 1955, 581.

2 Fred Winsor, "'Solving a Problem': Privatizing Worker's Compensation for Nova Scotia's Offshore Fishermen, 1926–1928," *Acadiensis* 18, no. 2 (Spring 1989): 94–110. Members of the Fishermen's Union of PEI in 1938 stated their opposition to the inclusion of fishermen in any Workmen's Compensation Act. Minute Book, 2 August 1938, Fishermen's Union of Prince Edward Island. Station No. 4 fonds: [1936–65] Acc 4346, PARO. Recent scholarship by Katie Mazer, "Making the Welfare State Work for Extraction: Poverty Policy as the Regulation of Labor and Land," *Annals of the American Association of Geographers* 109, no. 1 (January 2019): 18–34 takes on this theme directly.

3 William Schrank, "The Failure of Canadian Seasonal Fishermen's Unemployment Insurance Reform During the 1960s and 1970s," *Marine Policy* 22, no. 1 (1998): 67–81; William Schrank, "Benefiting Fishermen: Origins of Fishermen's Unemployment Insurance in Canada, 1935–1957," *Journal of Canadian Studies* 33, no.1 (Spring 1988): 61–87; Lund, "Fishing for Stamps."

4 Syda & Cousins (Digby) to Charles H. Tupper, 30 April 1894, RG23, FO, vol. 212, T-3129, File 1133, Part 1, Extension – Lobster Fishing Season, 1894–6, LAC. There is a considerable literature on outmigration Alan Brookes, "Out-Migration from the Maritime Provinces, 1860–1900: Some Preliminary Considerations," *Acadiensis* 5, no. 2 (Spring 1976): 26–56; Brookes, "The Golden Age and the Exodus: The Case of Canning, Kings County," *Acadiensis* 11, no. 1 (Autumn 1981): 57–82 and Patricia

A. Thornton, "The Problem of Out-Migration from Atlantic Canada, 1871–1921: A New Look," *Acadiensis* 15, no. 1 (Autumn 1985): 3–34 remain important.

5 See, for example, James Struthers, *No Fault of Their Own: Unemployment and the Canadian Welfare State, 1914–1941* (Toronto: University of Toronto Press, 1983).

6 Private email to author from Meg Stanley, 10 September 2021, citing "Report of Timber and Grazing Inspector, 2 July 1918," Banff Annual Reports, History Records, Parks Canada, Calgary Office.

7 Abram Hatfield to Prince, 26 February 1906, RG23, FO, vol. 297, T-3222, File 2314, Part: 8 Lobster Regulations, 1906–9; *Evidence Taken*, 1909, Prince, 43.

8 "Why the Lobster Fishery Industry Is of Prime Importance to the Maritimes," *Montreal Standard*, 27 October 1928, MG6 A, vol. 26, no. 1, Scrapbook of newspaper clippings etc. on the lobster industry, RS&Co., NSA.

9 "Fisherman" East Dover, NS, to Marine Department, 24 October 1910, RG23, FO, vol. 298, T-3981, File 2314, Part: 10A, Lobster Regulations – Minister's File, 1912, LAC.

10 Clifford Robinson to Prince, 5 July 1897, RG23, FO, vol. 213, T-3129, File 1133, Part 2, Extension – Lobster Fishing Season 1896–8, LAC.

11 Edouard Chiasson to Rodolphe Lemieux, 11 September 1903, RG23, FO, vol. 213, T-3130, File 1133, Part 4, 1903–5, Extension – Lobster Fishing Season 1898–1905, LAC.

12 J.S. Turbide PP, Camille Delaney & Aibin Thereault telegram to Préfontaine, 4 September 1905, RG23, FO, vol. 296, T-3222, File 2314, Part 7, Lobster Regulations, 1904–6, LAC. Reverend J.S. Turbide was a member of the 1903 Commission that examined fishing issues in the Bay of Fundy and the Magdalen Islands.

13 Préfontaine to Lemieux, 14 September 1905, RG23, FO, vol. 296, T-3222, File 2314, Part 7, Lobster Regulations, 1904–6, LAC.

14 Editor, "Says Special Lobster Season Paid," *CF*, February 1922, p. 35.

15 Rev. Father Alfred Boudreau, Arichat, 2 November 1927, 1436, Maclean Commission, MG6 A, vol. 18, no. 1 Canada. Evidence Vol. 5, Arichat, St. Peter's, Ingonish, North Sydney, NSA.

16 Gabarus, 28 June 1894, RG23, FO, vol. 212, T-3129, File 1133, Part 1, Extension – Lobster Fishing Season 1894–6, LAC.

17 Clipping, *Chronicle*, n.d. 1931, RS&Co., MG6 A, vol. 26, no. 1, Scrapbook of newspaper clippings etc. on the lobster industry, NSA.

18 "Annoyed with Rhodes' Stand Several Hundred Families Will Need Relief Says Lobstermen," *Chronicle*, September 1932, RS&Co., MG6 A, vol. 26, no. 1, Scrapbook of newspaper clippings etc. on the lobster industry, NSA.

19 *"Standard* Correspondent Informed of Continued Illegal Lobster Fishing," 26 September 1936, RS&Co., MG6 A, vol. 26, no. 1, Scrapbook of newspaper clippings etc. on the lobster industry, NSA.

20 H.W. Longworth to Senator Ferguson, 15 October 1895, RG23, FO, vol. 294, T-3219, File 2314, Part 1, Lobster Regulations, 1895–6, LAC; "Eastern Canadian Fishery Conference," *CF*, October 1921, p. 227; Canada, *House of Commons Debates*, 18.2.1, 26 January 1937 (Alfred Brooks, Conservative, Royal), 307.

21 "Lobster Situation 'Reasonably Satisfactory,'" *Herald*, 7 July 1934, scrapbook H.G. Connor, Fisheries Museum of the Atlantic, Lunenburg; *Second Annual Report of the Department of Fisheries 1931–2*, 13.

22 "Nova Scotia News," *CF*, November 1925, p. 347; Canada, *House of Commons Debates*, 16.2.1, 24 February 1928 (A.E. MacLean, Liberal, Prince), 783.

23 Canada, *House of Commons Debates*, 17.5.2, 13 March 1934 (James Ralston, Liberal, Shelburne-Queens), 2146.

24 Canada, *House of Commons Debates*, 17.5.2, 13 March 1934: 1443; 17.6.1, 6 February 1935 (James Ralston, Liberal, Shelburne-Queens), 547–9.

25 Canada, *House of Commons Debates*, 18.1.3, 28 April 1936 (Joseph Michaud, Liberal, Restigouche-Madawaska), 2271; "Fisheries Conditions 'Deplorable' Speakers Tell of Hardships Among the Fishermen of North Shore, Eastern NB," *Telegraph Journal*, 26 August 1938, p. 1. MG6 A, vol. 27, no. 6, RS&Co., Newspaper Clippings, NSA.

26 E.R. Forbes, "Cutting the Pie into Smaller Pieces: Matching Grants and Relief in the Maritimes in the 1930s," *Acadiensis* 17, no. 1 (Autumn 1987): 34–55.

27 Canada, *House of Commons Debates*, 18.4.1, 23 January 1939, 214, 217; 30 January 1939, 458 (R.E. Finn, Liberal, Halifax).

28 Canada, *House of Commons Debates*, 18.4.1, 2 February 1939 (Vincent-Joseph Pottier, Liberal, Shelburne-Yarmouth-Clare), 554.

29 Canada, *House of Commons Debates*, 18.4.1, 3 February 1939 (Finn), 459. 597.

30 Larkin, "Our Way of Living," 45.

31 James A. Brander and Gregor W. Smith, "Economic Research in Canada: Evolution and Convergence," *Canadian Journal of Economics/revue Canadienne D'économique* 50, no. 5 (December 2017): 1206; Doug Owram, *The Government Generation, Canadian Intellectuals and the State, 1900–1945* (Toronto: University of Toronto Press, 1986); Barry Ferguson, *Remaking Liberalism: The Intellectual Legacy of Adam Shortt, O.D. Skelton, W.C. Clark, and WA. Mackintosh, 1890–1925* (Montreal and Kingston: McGill-Queens University Press, 2014), 208; J.L. Granatstein, *The Ottawa Men: The Civil Service Mandarins, 1935–1957* (Toronto: Oxford University Press, 1982).

32 Wright, *A Fishery for Modern Times*, 43; Jennifer Hubbard, "Fisheries Biology and the Dismal Science: Economists and the Rational Exploitation

of Fisheries for Social Progress," in *Fisheries, Quota Management and Quota Transfer: Rationalization through Bio-economics*, ed. Gordon M. Winder (New York: Springer, 2018), 31–61.

33 Marion Fourcade, *Economists and Societies: Discipline and Profession in the United States, Britain, and France, 1890s to 1990s* (Princeton: Princeton University Press, 2009), 2, 8, 62.

34 Edward S. Mason and Thomas S. Lamont, "The Harvard Department of Economics from the Beginning to World War II," *Quarterly Journal of Economics* 97, no. 3 (August 1982): 383–433.

35 Banoub, *Fishing Measures*, 5.

36 Hubbard, *Science on the Scales*, 215; Hubbard, "In the Wake of Politics:" 373; more generally, see Hubbard, Wildish, and Stephenson, eds., *A Century of Maritime Science*; Hubbard, "Fisheries Biology and the Dismal Science."

37 Student Records, Steward Bates, Copy of Application for Commonwealth Fund Fellowship, Harvard University Archives, Pusey Library, Cambridge, MA.

38 "Obituary: Stewart Bates," *Trade News*, May 1964 (Department of Fisheries), 10.

39 Bates, *Canadian Atlantic Sea-Fishery*; See Stewart Bates to R. McGregor Dawson, 24 December 1943; Dawson to Bates, 6 July 1944, 2002–02/002 MS52.256.f.c.8, R.M. Dawson, Royal Commission Correspondence – Bates, Nova Scotia Royal Commission on Provincial Development and Rehabilitation Fonds, NSA; Miriam Carol Wright, "Fishing in Modern Times: Stewart Bates and the Modernization of the Canadian Atlantic Fishery," in *How Deep Is the Ocean?*, ed. Candow and Corbin, 199.

40 Wright, "Fishing in Modern Times," 195–8; and Wright, *Fishery for Modern Times*.

41 "Obituary: Stewart Bates," 10; "Stewart Bates, 56, Canadian Official," *New York Times*, 25 May 1964, p. 33.

42 Bates, *Canadian Atlantic Sea-Fishery*, 10.

43 Bates, *Canadian Atlantic Sea-Fishery*, 10, 159–60.

44 Bates, *Canadian Atlantic Sea-Fishery*, 65.

45 Bates, *Canadian Atlantic Sea-Fishery*, 10.

46 Bates, *Canadian Atlantic Sea-Fishery*, 65.

47 Bates, *Canadian Atlantic Sea-Fishery*, 84.

48 "Political Economy Club," *McGill Daily*, 25 November 1946; H. Scott Gordon, "Theorizing in Economics: a Study of Its Logical Foundations," (PhD diss., McGill University, 1953).

49 H. Scott Gordon, "The Fishing Industry of Prince Edward Island. A Report Prepared for the Prince Edward Island Development Committee" (Ottawa: Markets and Economic Service, Department of Fisheries, 1952)

published in *Report of the Prince Edward Island Fisheries Development Committee* (Charlottetown, 1952); Edward MacDonald and Boyde Beck, "Lines in the Water: Time and Place in a Fishery," in *Time and a Place: An Environmental History of Prince Edward Island*, edited by Edward MacDonald, Irené Novaczek, and Joshua MacFayden (Montreal & Kingston: McGill-Queen's University Press, 2016), 218–45.

50 Gordon, "Fishing Industry," 109; Gough, *Managing Canada's Fisheries*, 229.

51 Gordon, "Fishing Industry," 102–4, 105.

52 H. Scott Gordon papers, 1947–93, Archives online at Indiana University, http://webapp1.dlib.indiana.edu/findingaids/view?doc.view=entire _text&docId=InU-Ar-VAB8605; George W. Wilson, "Scott Gordon: An Appreciation," in *Welfare, Property Rights and Economic Policy: Essays and Tributes in Honour of H. Scott Gordon*, ed. Thomas K. Rymes (Montreal and Kingston: McGill-Queens University Press, 1991), 16-21; H. Scott Gordon, "The Economic Theory of a Common-Property Resource: The Fishery," *Journal of Political Economy* 62 (1954): 127, 128.

53 Gordon, "Economic Theory of a Common-Property Resource:" 132–4.

54 Gregory Ferguson-Cradler, "Liberalism in Numbers Only: Science, Politics and State Power in Postwar Global Fisheries Management," (PhD diss., Princeton University, 2016).

55 Hubbard, "In the Wake of Politics:" 373.

56 Carmel Finley, *All the Fish in the Sea: Maximum Sustainable Yield and the Failure of Fisheries Management* (Chicago: University of Chicago Press, 2011).

57 Hubbard, "In the Wake of Politics:" 371, 373; Hubbard "Fisheries Biology and the Dismal Science," 33.

58 Hubbard, "Fisheries Biology and the Dismal Science," 51.

59 Hubbard, "Fisheries Biology and the Dismal Science," 31.

60 Quoted by Gough, *Managing Canada's Fisheries*, 257.

61 Canada, *Department of Fisheries Annual Report 1956–57* (Ottawa: s.l., 1957), 31.

62 Canada, *Department of Fisheries Annual Report 1956–57*, 20.

63 Canada, *House of Commons Debates*, 21.3.1 31 August 1950 (Charles-Arthur Dumoulin Cannon, Îles-de-la-Madeleine, Liberal), 83.

64 Cabinet 15 October 1952, RG2, Privy Council Office, Series A-5-a, vol. 2651, T-2368, Fisheries; compensation to lobster fishermen for losses of gear in 1951 and 1952, LAC.

65 Cabinet 12 June 1953, RG2, Privy Council Office, Series A-5-a, vol. 2653, T-2369 Fisheries; losses of lobster traps in Prince Edward Island, LAC.

66 Cabinet 14 October 1953, 6 July 1953, RG2, Privy Council Office, Series A-5-a, vol. 2653, T-2369, Fisheries; amendment to Lobster Trap Indemnity Regulations, LAC.

67 "Lobster Trap and Boat Insurance Explained to Fishermen at Meeting," *Charlottetown Guardian*, 29 April 1954, p. 5; *Canada Year Book 1955*, 592.

68 J.L. Rutherford, D.G. Wilder, and F.C. Frick, *An Economic Appraisal of the Canadian Lobster Fishery* (Ottawa: Research Board of Canada, 1967), 54.
69 Cabinet 17 May 1966. RG2, Privy Council Office, Series A-5-a, vol. 6321, Assistance to lobster fishermen, Newfoundland, LAC.
70 Cabinet 25 January 1966, RG2, Privy Council Office, Series A-5-a, vol. 6321 Assistance to lobster fishermen, Newfoundland, LAC.
71 Jacquard, *Lobstering*, 301, 303–4.
72 L.S. Bradbury, Dir. Industrial Development Service Department of Fisheries to Louis Solomon, Solest Manufacturing, 17 April 1959, GN18/21, Department of Fisheries, Lobster Trap Designs and Designers, PANL.
73 Clipping, PEI report; "Methods Used to Introduce a New Type of Fishing Gear to PEI Lobster Fishermen," 34, GN18/21, Department of Fisheries, Lobster Trap Designs and Designers, PANL.
74 *Report of the Prince Edward Island Fisheries Development Committee*, 35, 38.
75 DeWolf, *The Lobster Fishery of the Maritime Provinces*, 26.
76 Jacquard, *Lobstering*, 376; Kathy Johnson, "A Look Inside Clearwater's Offshore Lobster Fishery," *Chronicle Herald*, 28 February 2019.
77 Schrank, "Benefiting Fishermen;" Lund, "Fishing for Stamps," 183; Canada, *Commission of Inquiry on Unemployment Insurance*, Part II (Ottawa: Minister of Supply and Services Canada, 1986), 241, accessed 29 July 2020, https://epe.lac-bac.gc.ca/100/200/301/pco-bcp/commissions-ef/forget1986-eng/forget1986-eng.htm.
78 Schrank, "Benefiting Fishermen:" 83, n. 28.
79 Wright, *Fishery for Modern Times*, 60.
80 Schrank, "The Failure:" 67.
81 Report of the Committee of Inquiry into the unemployment Insurance Act 1961–62 (Gill Committee), 110, 174, 177, accessed 29 July 2020, https://epe.lac-bac.gc.ca/100/200/301/pco-bcp/commissions-ef/gill1962-eng/gill1962-eng.htm.
82 Hubbard, *Science on the Scales*, 240.
83 Greg Marquis, "Confederation's Causalities: The Maritimer as a Problem in 1960s Toronto," *Acadiensis* 39, no. 1 (Winter/Spring 2010): 86; Historical Statistics of Canada, Section N: Fisheries Table N38–48 Number of persons engaged in primary fishing operations, by province, 1878 to 1975, accessed 29 July 2020, https://www150.statcan.gc.ca/n1/pub/11-516-x/sectionn/4057755-eng.htm; Kari Levitt, "Population Migration in the Atlantic Provinces," in *Canadian Political Science Association Conference on Statistics 1961: Papers*, ed. Wm. C. Hood and John A. Sawyer (Toronto: University of Toronto Press, 1963), 49–96.
84 Rutherford, Wilder, and Frick, *An Economic Appraisal*, 32.
85 "Fishermen Get Unemployment Insurance," *CF*, September 1956, pp. 2, 42.

86 Derek Johnson, "Merchants, the State and the Household: Continuity and Change in a 20ᵗʰ-Century Acadian Fishing Village," *Acadiensis* 24, no. 1 (Autumn 1999): 57–75.

87 Jacquard, *Lobstering*, 268.

88 "Consider the Lobster by Wayfarer," *Daily News*, 24 April 1962. GN18, Department of Fisheries, Departmental Files Box 48 32/4 Newspaper Clippings – Lobster Fishery, PANL.

89 Alexander Laidlaw Associate dir. Extension Dept to Loran E. Baker, 24 May 1956, RG30 3/1/ Extension Department, 4409–14 Fisheries, SFXUA.

90 A.A. MacDonald and W.B. Clare, "Guysborough Shore Resource Survey," Extension Dept 1966, p viii, RG30–3/1/ Extension Department, 6796, Fisheries, SFXUA.

91 MacDonald and Clare, "Guysborough," 98.

92 Hubbard, "Fisheries Biology and the Dismal Science," 52.

93 Rutherford, Wilder, and Frick, *An Economic Appraisal*, 7.

94 Agriculture and Resource Economics, accessed 3 August 2020, https:// saskaggrads.com/index.php?page=agriculture-resource-economics-the -early-years. Report of the Conference of FAO 1946, accessed 3 August 2020, www.fao.org/3/x5583E/x5583e04.htm.

95 Rutherford, Wilder, and Frick, *An Economic Appraisal*, 84, 85.

96 Gough, *Managing Canada's Fisheries*, 257–8.

97 Rutherford, Wilder, and Frick, *An Economic Appraisal*, 31, 96; Karen Foster "Productivism, Neoliberalism, and Responses to Regional Disparities in Canada: The Case of the Atlantic Canada Opportunities Agency," *Acadiensis* 48, no. 2 (Autumn 2019): 117–45. Foster notes the emphasis on productivity and growth since 1980: 118.

98 W.C. Mackenzie, "The Problems of the Lobster Fishery," *CF*, January 1965, p. 23.

99 W.C. Mackenzie, "How Can We Overcome Lobster Poaching," editorial *CF*, November 1960, pp. 7, 11.

100 Rutherford, Wilder, and Frick, *An Economic Appraisal*, 97; Flint, "The Lobster Fishery of Southwest Nova Scotia," 112; M. Patricia Marchak, "What Happens When Common Property Becomes Uncommon?" *BC Studies*, 80 (Winter 1988–9): 14.

101 Rutherford, Wilder, and Frick, *An Economic Appraisal*, 98.

102 Flint, "Lobster Fishery of Southwest Nova Scotia," 44.

103 Historical Statistics of Canada, Section N: Fisheries, Table N12–24. Quantities of fish landed, by region and by major species, 1869 to 1975, accessed 10 August 2020, https://www150.statcan.gc.ca/n1/pub/11 -516-x/sectionn/4057755-eng.htm.

104 D.A. MacLean, "Lobster Fishery Investigation, Pictou NS 1965," p. 4. Economics Branch, Department of Fisheries, Halifax, NS October 1965.

RG3–3/1 Extension Department, 711–7, Fisheries, SFXUA. Lobster fishermen in L'Archeveque and Forchu set voluntary limits before Lismore. Lobster Fishery Task Force, *Final Report*, March 1975, 65, accessed 26 May 2022, https://waves-vagues.dfo-mpo.gc.ca /Library/40628139.pdf.

105 Parsons, *Management of Marine Fisheries*, 171.

106 Gough, *Managing Canada's Fisheries*, 258.

107 Peter R. Sinclair, "Fishermen Divided: The Impact of Limited Entry Licensing in Northwest Newfoundland," *Human Organization* 42, no. 4 (Winter 1983): 307–13.

108 Gough, *Managing Canada's Fisheries*, 258–9; Lobster Fishery Task Force, *Final Report*, 65.

109 DFO, *A Brief History of the Lobster Fishery*, 22; Jacquard, *Lobstering*, 290.

110 Flint, "Lobster Fishery of Southwest Nova Scotia," 136; Marchak, "What Happens When Common Property Becomes Uncommon?:" 14.

111 Gough, *Managing Canada's Fisheries*, 284.

112 Matto Mildenberger, "The Tragedy of the *Tragedy of the Commons*," *Scientific American*, 23 April 2019.

113 Elinor Ostrom, "The Danger of Self-Evident Truths," *PS: Political Science and Politics* 33, no. 1 (March 2000): 37–8, accessed 21 May 2021, doi:10.2307/420774.

Conclusions and Epilogue

1 Bates, *Canadian Atlantic Sea-Fishery*, 10.

2 Gough, *Managing Canada's Fisheries*, 405, 409, 419.

3 "Seafisheries Landed Value by Province," 2017, accessed 23 May 2022, www.dfo-mpo.gc.ca/stats/commercial/land-debarq/sea-maritimes /s2017pv-eng.htm; Lobster Fishing Areas 27–38 Integrated Fisheries Management Plan, accessed 23 May 2022, www.dfo-mpo.gc.ca/fisheries -peches/ifmp-gmp/maritimes/2019/inshore-lobster-eng.html#toc3. Lobster was less than 6 per cent of the commercial landings for Newfoundland.

4 DFO, "Assessment of Lobster (*Homarus americanus*) in Lobster Fishing Area 34."

5 Giles Thériault, John Hanlon, and Lewis Creed, *Report of the Maritime Lobster Panel* (New Brunswick, Nova Scotia, and Prince Edward Island Governments, 2013), iii.

6 Gough, *Managing Canada's Fisheries*, 431; Paul Withers, "Canadian Lobster Exports Have Biggest Year Ever, Topping $3.2B Last Year," CBC News, 1 March 2022, www.cbc.ca/news/canada/nova-scotia/canadian -lobster-exports-topped-3-2-billion-last-year-1.6367872; Abe Streep, "Nova Scotia's Billion-Dollar Lobster Wars. How Indigenous Fishermen Are

Defending Their Rights – and Corporate Profits – in the Most Lucrative Fishery in North America," *New Yorker*, 21 May 2024, www.newyorker .com/news/dispatch/nova-scotias-billion-dollar-lobster-wars.

7   Eric Thomas and Leyla Sall, "Where Have All the Local Workers Gone? Reliance on Temporary Foreign Works in New Brunswick's Fish and Seafood Processing Industry," Economic Implications of Demographic Change in Atlantic Canada Conference, Halifax, 2–14, www.smu.ca /webfiles/ThomasandLallWherehavealthelocalworkersgone.pdf.

8   Flint, "The Lobster Fishery of Southwest Nova Scotia," 137; Parsons, *Management of Marine Fisheries*, 171.

9   Hubbard, "Fisheries Biology and the Dismal Science," 53.

10 DeWolf, *Lobster Fishery of the Maritime Provinces*, 1.

11 Lobster Fishery Task Force, *Final Report*, March 1975, accessed 23 May 2022, https://waves-vagues.dfo-mpo.gc.ca/Library/40628139.pdf.

12 "Restrictions Imposed on Lobster Licenses," *Globe and Mail*, 31 December 1975, p. B10.

13 Gough, *Managing Canada's Fisheries*, 349.

14 Gough, *Managing Canada's Fisheries*, 350.

15 Clotilde Bodiguel, "Fishermen Facing the Commercial Lobster Fishery Licensing Policy in the Canadian Maritime Provinces: Origins of Illegal Strategies, 1960–200," *Marine Policy* 26 (2002): 275; Fisheries Resource Conservation Council, *A Conservation Framework for Atlantic Lobster*, Report to the Minister of Fisheries and Oceans 1995 (Ottawa: Minister of Supply and Service Canada, 1995), 12; DFO, *A Brief History of the Lobster Fishery*, 31.

16 Allain J. Barnett, Robin A. Messenger, and Melanie G. Wiber, "Enacting and Contesting Neoliberalism in Fisheries: The Tragedy of Commodifying Lobster Access Rights in Southwest Nova Scotia," *Marine Policy* 60 (June 2017): 60–8. Licences clearly had value as a de facto transferable commodity, and through government buyback payments disbursements had a rapidly increasing market value.

17 Environment Canada, F&M Service, *Policy for Canada's Commercial Fisheries*, May 1976, 53, https://waves-vagues.dfo-mpo.gc.ca/Library/293.pdf.

18 *Policy for Canada's Commercial Fisheries*, 24.

19 *Policy for Canada's Commercial Fisheries*, 39.

20 McMullan and Perrier, "Lobster Poaching and the Ironies of Law Enforcement," 686.

21 Ralph Surette, "Liberals, Tories Fish for Votes," *Globe and Mail*, 3 February 1979, p. 8; Editorial "The Politics of Fish," *Globe and Mail*, 16 May 1979, p. 6.

22 Bodiguel, "Fishermen Facing the Commercial Lobster Fishery," 275.

23 Barnett, Messenger, and Wiber, "Enacting and Contesting Neoliberalism," 63.

24 Those who stayed in the fishery were likely to have taken on more debt as expenses grew faster than revenue. Now, the Fishermen Loan Boards

required fishermen to use their lobster licences (still not legally recognized as property!) as collateral on boat loans, giving the Boards the right to seize and sell both boats and licences in the cases of default in periods of high interest rates such as the early 1980s. The cost of fishing also increased with higher fuel prices. John L. McMullan, David C. Perrier, and Norman Okihiro, *Law, Regulation, and Illegality in the Nova Scotia Lobster Fishery* (Halifax: Atlantic Institute of Criminology, 1988), 19, www.publicsafety. gc.ca/lbrr/archives/ken%207747%20m3a%201988-eng.pdf; McMullan, Perrier, and Okihiro, "Regulation, Illegality and Social Conflict in the Nova Scotia Lobster Fishery:" 127. This is a slightly revised version of *Law, Regulation, and Illegality*.

25 McMullan, Perrier, and Okihiro, *Law, Regulation, and Illegality*, 2.
26 McMullan, Perrier, and Okihiro, *Law, Regulation, and Illegality*, 73; "Plant Hit in Protest over Lobster," *Globe and Mail*, 31 December 1976, p. 2; "RCMP Seize TV Film of Fishermen's Protest," *Globe and Mail*, 1 January 1977, p. 2.
27 Parker Barss Donham, "Tempers Flare in Lobster War," *Globe and Mail*, 16 June 1984, p. 8; "Irate NS Fishermen Burn, Sink Patrol Boat over Lobster Dispute," *Globe and Mail*, 12 May 1983, p. 1; Michael Clugston, "Nova Scotia's Lobster Wars," *Maclean's*, 28 May 1984.
28 McMullan, Perrier, and Okihiro, *Law, Regulation, Illegality*, 30.
29 Simone Poliandri, "Mi'kmaq People and Tradition: Indian Brook Lobster Fishing in St. Mary's Bay, Nova Scotia," *Papers of the Thirty-Fourth Algonquian Conference*, 34th ed. H.C. Wolfart (Winnipeg: University of Manitoba, 2003), 308.
30 This was misremembered as happening in the 1950s at Esgenoôpetitj, but lobster licences had no monetary value until after 1970 and the government programs for that region did not begin until 1978; Stuart Beaton, "Navigating Trouble Waters," *Navigator*, 16 October 2020; King, *Fishing in Contested Waters*, 106.
31 Thériault, Hanlon, and Creed, *Report of the Maritime Lobster Panel*, 80–1.
32 Caitlin Krause and Howard Ramos, "Sharing the Same Waters," *British Journal of Canadian Studies* 28, no. 1 (May 2015): 23–42.
33 King, *Fishing in Contested Waters*.
34 Linda Pannozzo, "United We Fish," *Halifax Examiner*, 24 November 2020. Online.
35 The frequently quoted numbers in Ken Coates, *The Marshall Decision at 20: Two Decades of Commercial Re-Empowerment of the Mi'kmaq and Maliseet* (Ottawa: Macdonald-Laurier Institute, 2019), 29, 33, for the value of lobster licences as up to $2 million appears to be an exaggeration.
36 Linda Pannozzo and Joan Baxter, "Lobster at a Crossroads: Part 3: What Are the Prospects for the Atlantic Lobster Fishery," *Halifax Examiner*, 10 October 2020. Online.

37 Brewer, "Governing the Fishing Commons," 167.

38 Joan Marshall, *Tides of Change on Grand Manan Island: Culture and Belonging in a Fishing Community* (Montreal and Kingston: McGill-Queen's University Press, 2008), 123.

39 Barnett, Messenger, and Wiber, "Enacting and Contesting"; "Regulations Amending Certain Regulations Made Under the Fisheries Act," *Canada Gazette*, Part I, vol. 153, no. 27 (6 July 2019).

40 Tina Comeau, "Stormy Lobster Waters Continue in St. Mary's Bay even After Teddy," *Vanguard*, 23 September 2020.

41 Linda Pannozzo and Joan Baxter, "Lobster at a Crossroads: Part 1: It's Been 20 Years Since the Marshall Decision, So Why Is There Still no Moderate Livelihood Fishery?" *Halifax Examiner*, 5 October 2020. Online.

42 Donald Silver Cameron, *Blood in the Water: A True Story of Revenge in the Maritimes* (Toronto: Penguin Random House, 2020); "The Lobster Wars of Bird Islands," *Halifax Examiner*, 26 July 2017. Online.

43 The important exception has been the serious reporting conducted by the *Halifax Examiner*.

44 Pannozzo and Baxter, "Lobster Fishery," Part 1.

45 Zoe Heaps Tennant, "The New Lobster Wars: Inside the Decades-Long Battle Between Fishers and the Federal Government over Mi'kmaw Treaty Rights," *Walrus*, 10 November 2020, https://thewalrus.ca/the-new -lobster-wars/.

46 Pannozzo and Baxter, "Lobster at a Crossroads: Part 1."

47 Richard Cuthbertson "Stakeouts and Microchipped Lobster: Inside DFO's Probe into a First Nations Fishery," CBC News, 5 October 2018, www.cbc .ca/news/canada/nova-scotia/dfo-first-nations-lobster-fishery -investigation-black-market-1.4846705; "Lobster Pound Owner Found Guilty of Illegally Selling Lobster," CBC News, 26 August 2020, www.cbc .ca/news/canada/nova-scotia/guang-da-international-sheng-ren-zheng -guilty-illegally-selling-lobster-1.5701123.

48 Tennant, "The New Lobster Wars."

49 Pannozzo and Baxter, "Lobster at a Crossroads: Part 3."

50 Jørgen Ole Bærenholdt and Nils Aarsæther, *Transforming the Local: Coping Strategies and Regional Policies* (Copenhagen: Nordic Council of Ministers 2001), 48.

51 Joan Baxter, "What About Clearwater and the Offshore Fishery?" *Halifax Examiner*, 6 November 2020; Linda Pannozzo, "In Search of Common Ground: An Interview About the Lobster Crisis in St. Mary's Bay," *Halifax Examiner*, 2 November 2020. Online.

52 Pannozzo, "In Search of Common Ground."

53 Livia Albeck-Ripka, "Climate Change Brought a Lobster Boom. Now It Could Cause a Bust," *New York Times*, 21 June 2018; Guilford, "The Enigma."

Index